LINEAGE

LINEAGE

LIFE AND LOVE AND SIX GENERATIONS IN CALIFORNIA WINE

BY STEVEN KENT MIRASSOU

VAL DE GRÂCE
BOOKS

Lineage: Life and Love and Six Generations in California Wine

First Edition 2021

Published by Val de Grace Books, Napa, California

Copyright © Steven Kent Mirassou

All rights reserved. No part of this book may be reproduced in any form without written permission from Steven Kent Mirassou.

ISBN 978-0-9848849-5-7

Library of Congress Control Number: 2021902507

Design by Connie Hwang Design

Printing by Artron Art Printing (HK) Ltd through Crash Paper

Photo credits: Jacket/cover, p. 21–22, 46, 58, 67, 82–83, 95–96, 120, and 143 by Ron Essex Photography; vii, viii-ix, p. 22, 28–32, 68, 144, 150, and 161–162 by Steven Kent Mirassou; xvi–p. 1 by Matt Toomey; p. 170 by Cindy Turchino.

∞

This book is dedicated to my growing family; I love you all. It is also dedicated to all of us who work in the hospitality business. We are compelled to take care of people, and there is glory in that.

∞

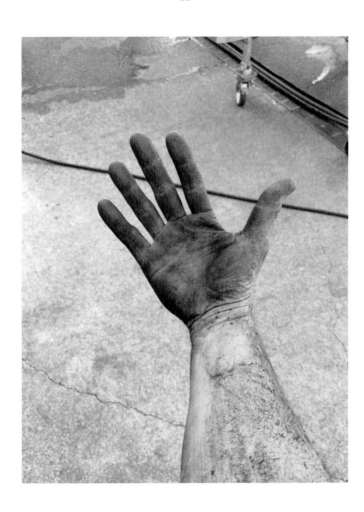

L to R: Edmund Mirassou, Norbert Mirassou, Max Heubner

∞

The Livermore Valley is located 49.4 miles east and south of downtown San Francisco, 26 miles east of the beneficent influence of the San Francisco Bay waters.

∞

CONTENTS

PART ONE

CHAPTER 1	The Deep and Simple Immensity	*3*
2	In the End, Bottomless	*11*
3	Every One of Them, Grand	*23*
4	Bone and Sinew	*33*
5	Making Wine is Morning Work	*47*
6	Real Cabernet Franc	*59*
7	Bleached and Silent and Solitary	*69*
8	Thanks You Can Never Repay	*77*

PART TWO

CHAPTER 9	The True North of Civilization	*85*
10	There is Something Sacred	*91*
11	Gods of Asphalt and Shingle	*97*
12	Like Birds Surprised into Flight	*107*
13	Imprisoned in These Barrels	*121*
14	Perfection is the Only Thing Left	*127*
15	A Clear and Melodious Song	*135*
16	Heightened Magic	*139*
17	Out From Land	*145*
18	Each Dusty Sunset Observed	*151*
19	She is Beautiful and Has Brown Skin	*163*
20	The Cleaving of Soul to Soul	*171*
21	Ageless Heart of the World	*175*
	Acknowledgments	*181*

PART One

CHAPTER 1

THE DEEP AND SIMPLE IMMENSITY

It is wondrously lonely out now. And it's cold in the night. Cold compared to the bed I came out of, untangling from the warm body of my woman, a half-hour before. There is not another soul anywhere around, and it is so dark here in the middle of my vineyard at three in the morning that I can only feel the vines, pregnant with Cabernet, surrounding me as my breath wreaths, like a phantom, away and up. I cannot see the ordered lines of vines progress away from me down the hill out to the horizon, but I know that they do. I cannot discern in the black the arroyos that cut through my property and put tears into the perfect rectangle of vines, but I know that they do and know that they are home to the fox and the coyote and the turkey when they are not running with the waters of winter. If one believes in God, he would find it here immanent in the perfect ripeness of fruit and in the gentle whisper of eucalyptus crown (that sounds like the sea coming to shore when the Livermore winds blow in the afternoon) and in the perfect coldness and the perfect blackness, made imperfect only by the perfection of stars salting the black bowl above.

In an hour the lights stationed on the perimeter of the vineyard will be lit as the generators are yanked to life, and the pickers will arrive from small towns to the east in their old cars and clothed for the chill. They will have their lugs and their knives, and they will amble to one of the rows that the foreman chooses, and they will start work there and work until there are no more grapes to pick. There is constant movement now as the pick is organized, one winery wanting fruit from some specific row so they start there; men and women hunch against the cold and to the height of the vines; the outhouse for the pickers will be towed to one of the lanes that separate one block of grapes from another, water coolers full of water will be set up by a truck tire and paper cups will litter the ground later like confetti at a parade. But this is the hour before the harvest starts, the time when I am still alone in this world where the blackness falls like soft rain and the cold burns like an electric arc.

The deep and simple immensity of this moment, here, in the still of the night-time world, will be inevitably attenuated by mundanity's ceaseless progress. In an hour, the trellis wires will thrum industriously in the silent air, hands will unmoor fruit unselfconsciously from vine and plastic lugs will be kicked ahead, keeping time to the tumbling bunch. That precious moment of quietude (both *wholesome* and whole) that reigned briefly before will be gone, lamented only by me and, maybe, only for that short while. The din will grow, pushing ahead of it into oblivion the plush silence, and the cold and shadowed spaces below the vines will be warmed by the moving lights and the pickers heaving their loads ahead.

The men and women who come into Livermore from out in the eastern valleys, their faces covered against the night, move in a graceless and determined knot from one row to another as, ahead, little ATVs haul the half-ton bins into which the smaller lugs are emptied. The vineyard rows they leave behind are diffident and tattered, bits of leaf lying bedeviled between rows. In the ever-moving lights, the flying bugs cavort, falling unbegrudgingly into the brightness. The artificial suns rip apart the sable fabric of the night, a dissonant juggernaut in their intractable singlemindedness. In the coming of the dawn the eastern rim of the foothills is set on fire, turning the black at the margins first to the blue of deep water, then bleaching the sky free of the black altogether. In the morning the vineyard is picked clean; the metallic melody of wire is silenced, and the grapes that gave these vines purpose have been hauled away to a purpose of their own. I stand here in the crystalline light, seeing every vine clearly now, laid out orderly, denuded, ~~running downhill~~, forlorn in their destitution. The vines have played out their predetermined role for another year and will die again for a while at the end of the fall. There had been a moment, though, in the black mercy of the night where all futures (for winegrape and winemaker) were still open out ahead and nature abided in grace.

.... ii ....

I remember the first time I saw Ghielmetti Estate Vineyard, back in 2005. It was late spring, and the green of the vines was like a slash of bright paint flung against the surrounding brown hills. The vineyard rolled east to west from 1000 feet to 600 feet above sea level; its rectangular shape complicated by two arroyos that gouged crooked lines in the regularity of the squared outline. The dried stream beds fold the surrounding interior spaces, so they do not true-up east-west and north-south everywhere. The off-center blocks ripen differently than the regular blocks then and add complexity to farming.

At that time, the vineyard was owned by a family business, good people, ultimately unreconstructed from a vision-perspective, though, and was planted, only as a cost of doing their main business. In our part of the Valley, if you built buildings, you had to set aside land for the growing of things. I don't think they'll ever realize that the vineyard they planted, that reached down deep and directly to the center of the earth, was the most valuable building that they did. The father and son, who planted the vineyard to offset their construction of a magnificent country club a few miles away, became partners of mine in The Steven Kent Winery. We determined, after months of meeting, that it made more sense for them to invest in a wine brand with demonstrated success and a desire to create world-class wines than it did to spend millions more on starting a new brand from scratch. The guarantee of control over the farming of an amazing vineyard and a sufficient supply of great fruit, we knew, could be a boon to the fortunes of the wine brand as well. So, we created a new company that combined the best of brand and vineyard that would provide amazing wines to our customers and put us at the forefront of quality wine in the Livermore Valley.

Ghielmetti Estate Vineyard is the site that carries the best of the Livermore appellation, the most inherent potential. It was laid out masterfully by the original farmers, the idiosyncrasies of variety and clone matched confidently to the lay of the land—above and below ground level. With the success of our wines, scores leading to increased prices for fruit, Ghielmetti Vineyard seemed on its way to becoming commercially successful on the high-end for our district. Each vintage, I would taste the wines made by other winemakers in Livermore from our fruit, and there was a distinctiveness to the site that shone through. Not each grape reflects the dirt in the same way, and the site is still very young, especially compared to great sites in Europe, but there is definitely a *Ghielmetti-ness* to the Sauvignon Blanc that comes off the site. In some years there are eight different winemakers making wine from this grape, and though styles differ—sometimes dramatically—there is a flavor and aromatic component that separates this site from others in Livermore Valley and California for that matter. Ghielmetti has become the jewel of the Valley.

················iii················

It has been a long night and I take off my long-sleeve shirt as the morning rapidly warms. I head to my truck thinking about fermenters and coffee. Next to the barn that lies in the middle of eight undeveloped acres at my vineyard there is a pile of hoops from old barrels that had gone into a fire the week before. The hoops are of uneven circumference, each reinforcing the bend of the stave at a different point along the barrel's length. The hoops of metal are pitted and stained, covered in mud in some places where they bit, on edge, into the wet earth. I've seen hoops made into sculptures in some places—abstract shapes or representing some animal or another. At Mirassou Vineyards, the family winery, long ago, someone set the smallest hoops inside successively larger ones at 90-degree angles until the whole thing resembled the orbit of atoms around the nucleus of an electric light bulb. The chandelier hung out on juxtaposed wires in an outside courtyard that served as a meeting place for large groups when the weather was nice. The ordered shape of that fixture caught my eye when I was a child running around the grounds surrounding the machinery of commerce. It was there too, I think, when I started to work at the winery in middle school. The chandelier had its place and seemed the natural bargain between utility and the rough-hewn imagination of one who puts working things together as a native consequence. There were a lot of things like this that I remember at the winery over the years, even some that survived until the closing of it in 2005. At the end, after the doors to the Tasting Room had been locked for the last time and the crush and fermentation equipment had been auctioned off to a group of buyers waiting, buzzard-like, for the sell-off to start, there was a pile of cabinets and wine racks and trestle tables and some old promotional signs and such. I don't remember seeing that light, though, at the end.

Matt Boughner, who managed the farming of Ghielmetti Vineyard for me until he left to grow pot in the Midwest, comes up to the truck as I am getting in and we talk about sugar levels and about the weather—always the weather. We plot the next blocks to come off—assuming we get the heat the forecast says we will get, but we've been fooled too many times to take that particular word as a sure bet. We shake hands, I thank him for his work, he trundles off to his truck, and I prepare to head to the winery, following the fruit. The day is alight as the sun rises triumphant and hot. The metal surfaces all around, the cars and trucks; the barn siding; house windows turn into mirrors that concentrate and focus and bring sweat up under the brims of hats.

The starlings flit in vines when they aren't dodging the talons of raptor, the rabbits run before trucks almost as sport, and the roads around the vineyard are clogged with cars taking the twists of the county roads over the nose-to-ass of the freeway. The morning explodes in activity and life. In the plenitude of the day, it is hard to imagine, unless you have been in it in the cold and clear darkness at harvest, that the exuberantly quiet night contains as much. In the mornings around harvest, chaos becomes the natural state. There is comfort in the cacophony because I can trace it back to the one thing we are all engaged in. No matter whether you are one of the migrant workers coming in from the small towns in the Central Valley in your old car to pick with your coats over long-sleeve shirts and bandanas warming the frigid air coming in in the night, or you are at the crush pad waiting to move that fruit along to the fermenter, or in the laboratory, in the tasting room, out on the streets showing samples, counting up the money coming in (rare) and going out (much less rare), you are all in service to the same heaving, gorgeous, messy beast.

This sweaty ferment that is the wine production business is like the snake attracted to its own tail, making eternal circles. Located in Northern California, as we are, where high technology dominates conversations and much of the realities of industry here, our work is the ancient, atavistic kind. Technology moves forward as it must, moving from one moment of invention to another, generally in a linear direction. But our business is connected to the oldest rhythms of the world, to the time at the very dawning of thought and understanding, when the discovery of wine (a product of chance to start) may have been responsible in part for the evolution of discovery to invention. While it is true that a bunchful of grapes, crushed a bit, and left to air, will yield wine without further intervention from man, this burbling liquid is to First Growth Bordeaux as a finger painting is to the Mona Lisa. 8,000 years of trial and error, the scholarly study of the finest Burgundian sites by monks more than 1,000 years ago, the cracking open of grape DNA leading to the naming of Cabernet Sauvignon's parents and Zinfandel's birthplace, may bring us all closer to an accounting of the *thingness* of wine; its intrinsic value, however, lies far outside the realm of the knowable. 8,000 years ago, a man or woman discovered one of the alchemical jewels endemic to wine: Time could be captured in the fermented juice of grapes. 8,000 years later, one of the greater gifts of wine is re-discovered, every vintage, by us as well. And so it is that as the timeline of my current harvest moves forward, it folds back as well, comfortably connecting me not only to all of my past seasons but also to every season that has ever come to pass.

My grandfather three times removed, 10 times removed, he, back to the first one, did what I am doing now at the same time of year I am doing it. He, too, stood in the middle of a pregnant, heaving vineyard in the middle of a frigid night, and thought too about the possibilities that existed in that pendant fruit until the moment that fruit was picked, and its singular and immutable course was set thus. This molecular connection to the previous iterations of myself stuns and comforts me still. I seek attachment to something larger than myself. I work to take care of people, to add richness and joy to their lives, to make the people who drink my wine a little happier after than they were before, each day. I am rooted to this purpose (though I did not realize this when I was a young man) and cannot do anything to change this course, even if I desired to. To those who are directed outward in their own search, my interior loop probably seems static, if not regressive. I would say to them that there is discovery to be made always, the furrow of my circular path cuts deep to be sure, but it also cuts wide.

The sun is behind me glowing as I head west to the winery. The few miles from the vineyard to the crush-pad take much longer than they should. New speed humps have been installed on the road, and I miss the window before the tech folks crowd onto Tesla Road to get to San Jose. As I turn into the driveway, I can hear the various mechanical rattlings associated with removing fruit from stem and getting the berries to the fermenters. On the scale, my fruit is being weighed. The half-ton bins will be slotted into the line of others that preceded them, and the berries will be removed from rachis and they will be crushed and fermented and shepherded into their new form. The new vintage has now begun, in the warm and dawning day, its inevitable journey, like every one of the harvests before it, all the way back down that infinite looping line. ○

"It is a quest to wring the best from myself, to lay myself out naked to the world with only my own sense of beauty and perfection to shield me, just as I necessarily must wring the most and best from the vines that delimit the finest pieces of dirt in my appellation."

CHAPTER 2

IN THE END, BOTTOMLESS

In place of comfort and certitude and all the other underpinnings necessary for the modern, rational man, there can be, instead, *magic*. If one is open to it, as I try to be, life is full of brief moments when doubt is transformed into revelation, when the mundane is cracked open like the shell of a nut and reveals a beauty that tilts the world a bit more on its axis and changes the way one conceives of meaning, when the consequences of past actions can be rewritten, and new paths open up to fresh and courageous exploration. To feel the sublime gravity of this non-mechanistic flow—this magic—under the veneer of modernity is to be wonderfully untethered from the ruthless causation of a life that ends up being not-quite-*lived*. It comes to me now, writing this on a quick airplane trip down to Phoenix, that a great glass of wine is the perfect medium to inspire this dive into the protoplasm of the possible. And that, perhaps, those who make it *are*, indeed, alchemists of a sort.

It's so easy to get lost in the trees when it comes to wine. Ron Rosenbaum, the author of *Shakespeare Wars*, has written beautifully about the bottomless nature of Shakespeare, about how each subsequent reading of *Hamlet* reveals something new and something deeper. I'd contend that this same immersive revisiting describes wine. Wine exists in a continuum; it starts life in a volatile state of heat and froth and ends it decades later (in the finest of cases) in a full-of-awe, breath-catching way. Wine grows and breathes and matures and senesces; it is a living object that can be at various times (depending upon serving temperature and bottle variation) as churlish as a two-year-old or as mannered as the charm school graduate. The great times of one's life—the sacraments, the ends of some beginnings, and the end of ends—are all presided over by wine.

There can certainly be an "inside baseball" aspect to all this, and I maintain that this interiority presages the very shape of the journey for those who are fated to dive deeply into depthless wine. I am incredibly fortunate to live this deep dive daily and to have a personal connection to all aspects of the process and to see the incredible beauty in the effort that goes in and the enjoyment that comes out that I cannot be unmoved by the details. Beauty has always been in short supply, and I work—in my medium—for its burgeoning.

To take in wine and to contemplate it as one tastes is to open oneself to a world of sensual mysteries that does not comply with rational rules. Parameters and perimeters have no measure here; time is fluid here, too: simultaneously, the before, the now, and the after. To taste a wine with one's heart is to come to know the dirt between the farmer's fingers, the exhalations of the vineyard as it sleeps in the black night. You will come to know also, the woman who takes the grapes from the vine with a practiced cut, the stooped man who rakes out the tanks after the juice is wine, and you will know finally the mind and the large heart of the exhausted winemaker who is filled to bursting when he gets the chance to take care of people.

Those enheartened, those true, who *feel* the wine, will not be alone. Each of them is connected with the all the others who feel too. These ephemeral filaments connect each of us across physical spaces and eons of time. Without groups of linked people there is no society, and there is no bond without that which exists in the contemplation of those things, such as wine, that reside on an emotional level. The rich Bordeaux, the earthy Brunello, the proper Burgundy are the shamanic keepers of the social fire around which we gather, all those susceptible and willing to come out of ourselves and become a part of a larger self, to give ourselves over to a vast underground sea of intuited and shared and unlonely experience. New experiences re-program the brain.

The process of considering anything deeply, be it great wine, fine food, book, movie, or your lover's naked hip is necessarily one of opening up, of being willing to jettison commonplace shapes in favor of those that put the lie to the contemptibly familiar. To taste some new bottle of wine, to slurp air through it in your mouth and feel it, to remark on the flavors and the way the wine makes your mouth feel, dryingly astringent or wetted by acid, and to let those sensations rebirth past experience is to insist upon living an un-ordinary moment. No more, then, does one accept perfunctory ritual. We begin to question our ordinary habits, and our personal growth begins as we unmoor ourselves from our previous expectations. Perhaps this unmooring leads only to a brief moment of recalibration, a moment of window-opening that lasts but a blink—providing a breath of cleansing air—quickly folding back upon itself and shuttering. Or, instead, it may thrust that window open for all time, emphatically and irrevocably connecting us to this new and shared ethos. It is no coincidence that Matisse's windows are always open.

To be open is to be subject to the vagaries of life… the unkind, the cruel, even. Open to the magical moments, also. It is the optimistic ones that I admire most. The cynical ones, in addition to not knowing the value of the thing, vaguely sense only the surface of it. The optimistic ones are stronger than the naysayers. They know that their hope for the future, the essential goodness of things, is hard won. It is much easier to be pleasantly surprised by positive outcomes when one continually expects the opposite than it is to perpetually overcome the disappointment in the failures that one trusts are but the mere exception. I number myself among the former and would rather be taken for a credulous fool than to be one who refuses to take a bite of life's bitter-sweet apple.

I realize that tilting at windmills is probably a buried need to have an excuse when things don't work out. Deep down, maybe true. I tend to look at such struggles, though, as usefully noble acts. I am attempting to make one of the world's finest wines—a blend of the five historic Bordeaux varieties—from my estate vineyards in the Livermore Valley. This endeavor is not just a labor of passion and love; it is also a labor of pissed-offed-ness, of pride, of envelope-pushing. It is a quest to wring the best from myself, to lay myself out naked to the world with only my own sense of beauty and perfection to shield me, just as I necessarily must wring the most and best from the vines that delimit the finest pieces of dirt in my appellation.

The Livermore Valley is located 49.4 miles east and south of downtown San Francisco, 26 miles east of the beneficent influence of the San Francisco Bay waters; 32.5 miles north and east of San Jose, and 40.2 miles east and south of Oakland. It is one of the last true Bay Area appellations. The Livermore Valley is also one of the oldest wine-growing districts in California. Grapes were planted there in the 1840s, about the same time as in the Napa Valley, and like its famous neighbor to the north, it is graced with warm growing-season days, an abundance of different soil types, micro-climates, history, and passion for producing wines of world-class quality.

Like much of California in the mid-1800s, the territory that would encompass the town of Livermore was divided up into enormous tracts and administered by the Mexican government. The Mexicans had succeeded in kicking the Spanish out of California in 1822. Closest to Livermore, Mission San Jose was secularized, along with all the other California missions in 1833, and the Catholic church lost much of its authority, though its creed carried on in the forcibly converted Ohlone and Costanoan indigenous peoples and in the Mexican settlers moving up north with the empire.

Mexican governors, far removed from Mexico City, 2,100 miles south, carried on the business of the country in Alta California. About ten years before the discovery of gold in the streams of the Sierra, foreign nationals, with relationships formed from convenience and mutual aggrandizement, were able to acquire large swaths of native land. Robert Livermore, an English sailor, jumped ship in Monterey and made his way north to what would be called Livermore in 1839.

Livermore befriended the local governor, and after converting to Catholicism, he was granted a plot of land measuring 48,000 acres. The erstwhile sailor found solace and riches away from the sea, opening up a hotel and a variety of other businesses just in time for the gigantic influx of fortune seekers chasing gold. Livermore died in 1858, but the unincorporated area in which he set up a number of hospitality-oriented businesses was named for him by William Mendenhall, the founder of the town of Livermore, in 1869. As the town grew, and men like Charles Wetmore, the first agricultural commissioner of California, envisioned the Valley as a prime place to grow grapes, its reputation for quality steadily increased. His winery, Cresta Blanca, produced the Semillon-Sauvignon Blanc blend that won the grand prize for the finest white wine in the world, out of thousands of entries, at the Paris Exhibition of 1889.

Before Prohibition there were about 50 wineries in Livermore. It took nearly eighty years after Repeal for the valley to regain that number. Much of the suburbanization that ended up isolating Livermore in the middle of a sea of freeway and house and diminished the concept of the Livermore Valley as the idyllic, world-class growing area by which it had been identified, before the mid-20th century, can be attributed to the creation, from a moribund U.S. Naval air station decommissioned after World War II, of two world-class laboratories. The Lawrence Livermore National and Sandia labs employ some of the smartest people and highest tech in the world. From a sleepy, agrarian town of 4,000 inhabitants before 1940, Livermore's population increased dramatically to over 80,000 people in the 50 years to follow. The labs also spawned a considerable number of family wineries over the years. The first wave of winery owners from the 1980s and 90s included a handful of former lab scientists that had settled in Livermore decades earlier, had acquired property, and had devoted part of their lives to the growing of grapes and the making of wine. Even though the reliance on the products and the lifestyle of a pastoral, frontier town (Livermore is the Bay Area's easternmost city) began to give way after the second great war, the bifurcation into two main industries (wine and weapons) was arguably completed some thirty years later.

The Livermore Valley can be a frustrating place to make wine. It is closer to the eight million people of the greater Bay Area than practically any other wine appellation, its downtown is finally flourishing, more progress is expected to come with the construction of an upscale hotel;

its viticultural chops are inarguable with wineries routinely receiving 90+ point scores in the press (my brands, Lineage Wine Company and The Steven Kent Winery, received 100-point scores by an esteemed wine critic in 2020) yet it lags behind in recognition regions far less able to make compelling wines. Livermore has been, up to this point, immune to the pressure of making great wine. Livermore does not yet exude a sense of the special or impart a challenge to the visiting wine drinker to expect world-class levels of vinous and experiential quality. Very few of us (I am, decidedly, one of them) believe we can equal the quality of more famous appellations so there is a natural cover for not putting the effort or the investment of dollars forward to prove our detractors (such as they exist) wrong. Once one subscribes to the notion that he has topped out on his quality pursuit, he can rationalize just about anything. Rationalize the use of older barrels, rationalize the purchase of fruit from a mediocrely farmed vineyard, rationalize the trickling stream of visitors. This rationalization finds, as a matter of course, the subjective answer to the question it seeks.

As one of the longest-tenured winemakers working in Livermore, and one of those (along with Karl Wente, David Kent, and Lori Souza of Wente Vineyards, Darcie Kent Winery, and the Tri-Valley Conservancy, respectively), most convinced of the region's world-class viticultural potential, I have spent a great deal of time thinking about the disconnection between the inherent quality of the Livermore Valley and the paucity of attention we have received. One of the chief challenges facing those who are serious, quality-driven winemakers here is the lack of an active, cooperative, and unifying vision among all the major wineries of what our community is capable of becoming. The wine business is part farming, and part hospitality, two industries that are always seemingly on the brink of ruin, especially in times of dramatic climate change and global pandemic. The wine business is also capital-intensive, emotionally and physically demanding, and unkind to those who do not take the challenges seriously. When there is an opportunity for like-minded people who are facing the same challenges every year to get together to help lift each other's fortunes by improving the quality of their products, it seems like rallying around a message of world-class quality would be an obvious first step to inoculating ourselves against many of the subjective vagaries of our industry. People being people, however, and individual business plans being non-duplicatable, we cannot always easily get to the obvious answer.

We are in the process of creating, however, a mechanism by which those wineries who are really serious can band together to show ongoing progress toward world-class quality to the consumer, critic, and wineshop owner. The *Mt. Diablo Highlands Wine Quality Alliance*, known as the MDHWQA, is an organization whose members will submit their wines to independent third parties for technical and qualitative evaluation with the goal of raising the quality of all the vineyards and wineries in the Valley. As one might imagine, laying oneself out (through one's wines) for a *go/no go* verdict by a stranger is like submitting one's offspring to an *is-your-baby-ugly?* contest. It is difficult to give up control of the quality narrative before the wine ever hits the public or the press. But third-party evaluation is like blind tasting; there isn't a better way (if the label and the putative prestige attached to it are hidden) to gain an honest, objective consideration of the wine qua wine. Moreover, if a third-party deems the wine to be of high-enough quality to carry the imprimatur of the appellation, the consumer gets better wine, the grape grower can charge higher prices for fruit (and, perhaps, make enough money to replant vineyards), and the winery can charge more for the bottle (hopefully insuring the continuation of the family business). Win-win-win.

The MDHWQA is in its infancy, and we will see over the years whether our thesis about the relationship of quality to the growth of recognition and more financial stability comes to pass. The Livermore Valley consists of my brands, The Steven Kent Winery, Lineage Wine Company, L'Autre Côte Wines, and Mia Nipote, as well as 3 Steves Winery, Arroyo Cellars, Bent Creek Winery, Big White House Winery, Boaventura de Caires Winery, Caddis Winery, Cedar Mountain Winery & Port Works, Charles R Vineyards & Winery, Concannon Winery, Crooked Vine Winery, Cuda Ridge Wines, Dante Robere Winery, Darcie Kent Vineyards, Eagle Ridge Vineyard, Ehrenberg Cellars, Elliston Vineyards, El Sol Winery, Fenestra, Garré Vineyard & Winery, Las Positas Vineyards, Leisure Street Winery, Longevity Wines, McGrail Vineyards & Winery, Mitchell Katz Winery, Murrieta's Well, Nella Terra Cellars, Nottingham Cellars, Omega Road Winery, Page Mill Winery, Retzlaff Vineyards & Estate Winery, Rios-Lovell Winery,

Rodrigue-Molyneaux Winery, Rosa Fierro Cellars, Ruby Hill Winery, The Singing Winemaker, Wente Vineyards, and Wood Family Vineyards.

Not all of these establishments will join the MDHWQA, though they will have the opportunity to do so. Not all will make great wine, and not all will survive long-term. But those that do become members and do make the commitment to world-class quality will have a much greater opportunity to have a beneficial impact on their bottom-lines as well as to improve the experiences of those who come to our valley.

The how-is-Livermore-different-than-Napa question has little to do with inherent grape-growing ability and practically everything to do with vision and money and romance and brand awareness. The fact that the first international gold medal awarded to a California wine was for Cresta Blanca's Livermore Valley white blend and the fact that there were more acres planted to Bordeaux varieties even into the 1960s in the Livermore Valley than in Napa has absolutely no value in this day and marketplace. Except for this one: the proof of concept has already taken place. There is nothing inherently inferior about my valley. I should be able to make a wine of exquisite distinction and palate-dropping quality. This is an edifying realization: it is all up to me. I have to grow great fruit; I have to choose the right barrels and make the right decisions in the cellar. I have to tell the right story (this is probably the hardest, and maybe the most important to do); I have to convince the press and the right customers; build a world-wide distribution network; create the right packaging... and do it every year for 10 or more years before anybody pays attention. Now you see my fondness for windmills... or Augean stables.

The Livermore Valley—and every other region in California—toils in the shadow of Napa. Napa is a beautiful place, thousands of acres of vineyards, great restaurants, lots of talented people... even more money. The one thing it isn't though is *de facto* the best place in California to grow Cabernet. Some would rather be lucky than good; most others would rather be pushed along on the uncomplicated wave that is inertia instead of making their own headway. Napa acquired the stature it did on its own merit for the most part. The visionaries migrated there—Mondavi, Brother Timothy, John Daniel, among a few others—and the place was cocooned far enough out of the way that it could ultimately conform to the non-historical, Edenic vision of these few men. Influenced though unbound by centuries of European oenological history and the plethora of traditions, varietal and other, and the regulations attending them, Napa became a monoculture of wine with awesome speed. Its hegemony was helped immeasurably (as was California wine to a lesser degree) by the Paris tasting in 1976. I think it was about this time and through the first part of the 1980s that Napa was really separated from the rest of California wine; I also think this was mostly a phenomenon of the nascent wine press.

Napa was easy to understand. It was a relatively small area doing one thing well; there were a lot of talented folks moving into the area in the early 70s and even more in the 80s and 90s, and the monied interests followed soon after the pioneers. Restaurants and hotels followed successful wineries, and it did not take long for all the elements of successful hospitality to be present in one small place at the same time. The virtuous circle was, indeed, closed when the immensely powerful wine press (Robert Parker and the *Wine Spectator*, foremost among them) found in the Napa Valley "style" a perfect match between the vinous object and their conception of its perfectibility. The wine lover with a cellar and the financial means made their way, then, to the 40-mile-long valley and made it one of the greatest wine success stories in history. The dynamics of wine-culture creation (some accidental, some the result of the accretion of small advances, some the result of hard work) created a zero-sum game for other wine regions, like the Livermore Valley. You can't have more than one place thus anointed, especially if you are trying to peddle the same variety. Everyone else picks up the crumbs and makes the best of it.

The greatest wine in the world is only as good, ultimately, as the number of people who are willing to pay for it. Quality does not exist in a vacuum; it is the sum of word-of-mouth, old-fashioned luck, persistence of communication, purity of vision, and purity of fruit (among an uncountable number of other factors). The fact that we have labored so long in relative obscurity is meaningful in a negative fashion only if we did not love, so much, this fucking business. This business of trying to translate the testicle-shrinking unreason of Mother Nature into something of transcendent beauty is not for everyone, to be sure; it IS for those that are essentially optimistic even against the worst odds. And for those who see beauty in imperfection, this business of fine-wine production, this business of memory creation, dealing in a currency of feeling, provides lifetimes of challenge, lifetimes of diving into the depths of an emotional enterprise that cannot be sounded, and that is, in the end, bottomless. o

CHAPTER 3

EVERY ONE OF THEM, GRAND

The steam from the hot coffee in the cold air of the morning sends up plumes. My wife, June, and I are out on the patio early in the morning watching, bit by bit, the vineyard appear out of the morning dark below as the sun climbs the eastern sky behind us. It is fall, and much of the fruit has been harvested already, but the blocks on this side of Ghielmetti Estate Vineyard have a few more days of hang-time on the vine ahead of them.

It is quiet out, except for the dogs rooting around, impatient for their breakfast; I'm rolling the hot mug in my hands to warm them up. The two of us talk together a lot, but we are quiet now. Words are a poor stand-in for the beauty and amity of this place. June is scrunched down in her Adirondack chair with a beanie shaped like a tiger (a gift from her brother) on her head to ward off the chill. She sits there, still and connected and rooted deeply, ticking to her own internal timepiece. June was like that when she was 18 (I was 18, too, meeting her the first day of college), and she is like that now more than 30 years later, here, 3,000 miles from her first home in the middle of a vineyard in the cold morning, meditating on the far margins. I go into the house and come back with the coffee pot to top us off. The dogs are barking at a rabbit at the fenceline when I return; the peace is gone now, and June is rising from her chair slowly to move back into the warmth of the house.

I remember wanting to go to school on the East Coast because it was far from home. Several generations of my family had been born and lived their lives in San Jose, where the family winery was, and it seemed peculiar to me to never have ventured away. I ended up at The George Washington University in Washington, D.C., and lived my first two years in a dorm three blocks from the White House. We'd play intramural football on the Mall in the shadows of the Lincoln Memorial, and we'd be pushed away by the Secret Service as the President's helicopter landed on our 50-yard line. I was far, far away from the vines and the fermenters.

Walking down the hallway of your freshman dorm on your first day, looking for your room, the doors are flung open everywhere, voluble conversations are batting about the ceiling and mingling along the fire sprinkler pipes; music of all types is spilling out from cassette players into the corridor, kids are running back and forth from room to room, and beautiful girls are all over, wearing summer shorts and tight, torn-at-the-neck t-shirts, speaking in accents and in other languages. I don't think I knew before I arrived that our dorm was co-ed; I have never experienced, since then, a purer sense of sexual yearning as I did that first day in Thurston Hall.

I lived in Room 504 that first year and had three roommates, one of whom was my best friend from home. On the door of each room, there was a placard that listed the names and hometowns of each of the occupants. We had two from California, and one each from Florida and Pennsylvania. I threw my stuff in my room and walked around the halls, ducking into rooms and introducing myself. In 507, there were three girls, all from the East Coast. They were all so foreign to me, so much more self-possessed than I—inhabitants of, and not strangers in, this new place. One of the girls was June from New Jersey. I saw her name and hometown on her door and didn't think I had ever met a June, except for Mrs. Boudreau, a woman who always seemed to be older than she was and who never could meet you eye-to-eye, who lived (next) door to the house I grew up in San Jose. A week or so after I moved in there was a party on the floor of the dorm. I cannot remember the theme; it might have been *Pink or Kink* or something like that. I do remember drinking beer and a dark room with green strobe lights and loud music; I remember, too, running into June there, and what she was wearing and talking to her for hours and connecting and feeling so outside my element, being undone by her self-possession and by my very innocence. I walked her back to her room that night and tried to kiss her in the dwindling and hopeful hours, and she closed the door softly in my face. I was married to that girl for 26 years, raised four children with her, learned from and was inspired by her, went through all the hardships and challenges and glories a couple can have in the quarter-century before I held her hand as she died, and I still don't know why she closed the door in my face all those years before, lips untouched.

June dated my best friend from home for most of college, got pregnant after graduation and had an amazing kid—April. June and my boyhood friend were not destined for a long life together, and after I returned to the East Coast for graduate school, having taken a year off, June and April and I started dating. I don't know how love works or whether it's even knowable, but a fortuitous college-dorm meeting grew into a death-do-you-part, forever, marriage.

June would drive to my apartment in New York and I would take the train from New York to the Trenton stop in New Jersey (her folks lived in Washington Crossing, PA, just over the Delaware river, and June would pick me up, April strapped into the car seat in the back) and we married a couple of years later. I adopted April, though that was just a formality as I am the only father she's ever known. I worked for a software company in lower Manhattan for a couple of years and rode two trains into work each morning, arriving ultimately at the World Trade Center, from which I would walk 10 blocks north to my office. My future did not reside in customer service or in computer software, and I did not envision another three years of study and dissertation writing to get my Ph.D. so we packed up our belongings, rented a U-Haul and drove 3,000 miles in eight days across the country until we decamped, finally, at a double-wide trailer in San Jose. Over the next two years we had the last two kids, Katherine (who was born in that double-wide, and thought, when she was about seven years old, that she would be the last U.S. President to be born in trailer) and Sara, born in a hospital in Morgan Hill, 45 minutes to the south, nineteen months later.

Our early life together was circumscribed by babies, trying to make ends meet in the Bay Area as young parents, and figuring out the future.

We had very little time alone together for the first seven or eight years of our marriage, but we did—because of the generosity of my parents-in-law who would come from Pennsylvania to watch the grandkids—get a week in Key West for our anniversary for a while. The early days were hard, especially when I started traveling for work every other week, and June had four kids to care for, mostly by herself. The end-of-week reunions when I'd fly back into San Jose were sweet for me, but I think there was an emotional withdrawal made over time, because of the nature of the work, that negatively affected our relationship for many years. When Sara was around seven, June went back to school to get her teaching credential, and she would spend the next ten years or so teaching the fifth grade at the elementary school around the corner from our house.

When the four kids were out of the house in 2014, we began finalizing plans to move to Livermore. We spoke intermittently about moving up to Livermore for years, but the time was never right, especially as the kids got older and more attached to their friends and their schools. The clearer it became that our financial security was wound up with the success of our business, it made sense to be closer to the center of our wine life. It did not hurt, of course, that the Livermore Valley is an enchanted place, full of wonder and magic. June was ready to take the next step in her life, ready to look at the family business as more than an adjunct to her marital experience, something she could actively help build. We had not found the right place to live at that point, but we were both attracted to the vision of a family estate, a place where our grandchildren could visit and run free and learn about the natural world at their pace, driven by their own senses of independence and curiosity and where our dogs could become again the untethered beasts that were hidden in their DNA.

Like many couples who met when they were young and who spent most of their lives together, we would occasionally revisit our origin story and all the lucky breaks and inevitabilities that led to this long life together. June and I would talk sometimes about the idea of soul mates, and I'm not sure where we actually came down on the subject. I do know, though, that all the time we were married I saw our relationship, the way we fit together, as inevitable in every positive way. I was fortunate to meet this person at a time in my life when we were both available; our tabs and slots matched, and I was perfectly happy to spend all the days I'd have with her. I did, in fact. But they were all the days she had with me, instead. They weren't nearly enough; they were though, every one of them, grand. ○

{ GENERATION I }
Pierre & Henriette Renaud Pellier
(Steven Kent Mirassou's great-great-great-grandparents)

∞

{ GENERATION II }
Pierre & Henriette Pellier Mirassou
(Steven Kent Mirassou's great-great grandparents)

∞

{ GENERATION III }
Peter & Justine Schreiber Mirassou
(Steven Kent Mirassou's great-grandparents)

∞

{ GENERATION IV }

L. Norbert Charles Mirassou; *R.* Edmund Aloysius Mirassou

(Steven Kent Mirassou's grandfather and great-uncle)

∞

∞

{ GENERATION IV, V, & VI }

L to R: **Norbert Mirassou** (Steven Kent Mirassou's grandfather), **Steve Mirassou** (Steven Kent Mirassou's father), **Steven Kent Mirassou** on the occasion of the celebration of the family's 125th anniversary in California wine 1979.

∞

CHAPTER

BONE AND SINEW

My family has been making wine in the U.S. longer than any other, seven generations—165 years, now that my son, Aidan has become our Assistant Winemaker. Every time I walk into my tasting room, I see the stolid countenances of the early generations of the family posed and stiff and formal. And these pictures have traveled, like treasured keepsakes, from our first tasting rooms at Mirassou, then to the re-invented brand under Gallo ownership where they were just curios in a marketing campaign, and then onto my walls, the rightful heir. As often as I had seen those sepia-ed photos when I was a kid running around the fermenters and disgorging champagne, the family legacy was never truly a part of my bone and sinew until I made my first vintage of wine nearly 20 years later. And 25 years after that, with the next generation working alongside me, I know now that I will be on that wall at some point too, and that my issue—more generations of our singular lineage—will view that old man and wonder, why the sly smile?

With each passing harvest, each fermenter dragged clean, vintage after vintage; each wine resting in barrel, growing in roundness, growing in excellence; each season tasting the wines of my team's labor, I realize, many years into my journey, that I had met a vocation commensurate to my capacity for idealism. More than simply the pursuit of personal excellence, I have come to understand that my time in my work, however long it should last, will be but one circle nested inside larger concentric rings. There is the circle of my family, and its role in California wine. My great-great-great grandfather is reputed to have been the first to bring Pinot Noir and Mourvèdre to California in 1854. He was certainly the first to bring the French prune, its shriveled fruit laying the cornerstone for a huge industry in Santa Clara County over the next century-and-a-half. The fortunes of the family waxed for a short time until they didn't anymore, and it falls to me and my son to redeem what had not been fully realized.

California is also a ring. The state conjures up for many a sun-soaked vinous Utopia, home to voluble, ebullient wines. Its earliest efforts are known only to a few, being too young and too far removed from the European centers of wine to drop much of a pebble in the larger pond. The origin story of California presaged many of the booms and busts the state would experience after the gold rush in 1848 created it; its wines were only curiosities at the margins of an old empire until the Paris Tasting in 1976 proved to be a rush of its own, rolling out at nearly the same pace.

The greatest concentricity is the story of wine itself. Wine is the story of wars, the story of and by the poets; it is the harbinger of culture, the loosener of tongues; it celebrates births and memorializes those gone; wine is a multi-billion-dollar business, undertaken mostly by tens of thousands of us who can barely pay our bills. Wine ultimately connects all of us who drink, and binds us all, through its vineyards, to the very beating heart of the green world.

.... ii ....

My son, Aidan, is one of the more wonderful people I know. Strong and sensitive, open to the possibilities of the world, if not always eager to search for them. He started out in our tasting room when he was a teenager, polishing glasses, emptying dump-buckets, throwing out the garbage, the same mundane but needed tasks that I did at his age, though mine took the form of warehouse and bottling-line work. My strategy for sharing generational progress was to show him as much of my world as possible, to put the best wines possible in front of him, to share my enthusiasm with him, to take him into the cellar to see what went on out of view of the customer. I hoped he would catch on, would be as awed as I was at the thought and act of making something beautiful.

His first couple of years in the cellar were a continuation of the scut work he did in the tasting room before he began to be looked upon as a potential heir there, not the boss's kid, but one who would create the paths forward. Winemaking seems like high living but only to those who haven't really seen it. It's mostly about being clean. There are a lot of bugs in a winery. Some are benevolent and bring about the fundamental and magical transformation of juice to wine, and some seek only to loose the microscopic dogs of war upon our otherwise civilized endeavor. It is the winemaker's job (really, the cellar rat's) to make sure as few havoc-wreakers as possible make contact with the stuff we'll eventually drink. So, Aidan washed a lot of fermentation bins, washed a lot of pumps, washed a lot of buckets and beakers and tanks and presses. One of the points you're trying to slam home is the tenuousness of this thing that we do, the fragility of the wine, its susceptibility, left unwatched, to run to vinegar. One of the best ways to illustrate this is, I think, to believe in the theory of germs as one would believe in the Old Testament holy father...to take it on faith that one moment of un-virtue leads to death. The kid survived his baptism quite nicely. His room may be a fucking mess, but you can eat off his fermenters.

Within our cabal, the woebegone life of the cellar rat is really one of great dignity. The person who knows the value of order-following and clean-making is held in high esteem. Aidan was this person then and is to a much greater degree that self-same person now. Beth Refsnider, another Assistant Winemaker, hired in 2018 as a production assistant, has assumed this role presently, and she has the makings of greatness. Attuned to nuances, aggressively searching out the work and the tasting opportunities, she will have control of her destiny in this business.

As each season rolled on, Aidan was given more responsibility, introduced more and more to the esthetic side of the craft, and given more and more a window into the whys of what we do. In 2017, my assistant winemaker, at the time, left to take a similar position with another winery in town, and Aidan stepped up to take over that set of responsibilities, that he wanted the job and was ready to become one of the initiates into our glorious and bedraggled order. He has embarked on his path, learning from his father, discarding those things that do not ring true to him, learning and doing, aspiring and fulfilling, perhaps in some Freudian way killing the father to lay with the dame, wine. Around the circumference of his ring or along the next knot on his line, over the years, from one harvest to another, he will draw out his own chord of the family lineage.

iii

The ramifications of my son's evolution didn't hit me at the moment he came into the lab to tell me he wanted the job. I was, selfishly, trying to figure out how much work this new reality would add to my life and how I was going to be able to handle it, how I was going to be able to be at the crush pad 24/7 at the same time I was supposed to be selling the wine that kept the company floating. It was a bit later I realized that I had been given a great gift, one I'm sure he didn't contemplate nor understand then. To be able to wage war against the lack of ambition and imagination that seems to permeate the winemaking community of Livermore, to build a monument to beauty and joy and richness, and to do it with your son is a deeply profound experience. Our wine family is the oldest in America and older than many that ply these same waters in Europe, and it is far likelier that a succeeding generation is going to shove off to find another lane, to make a different life for itself, than it is to stay to create something monumental from the pieces the father baled together.

We are on a quest, one as selfish and selfless as any of the famous ones of story and song. There is no guarantee of success, no guarantee that the next vintage might not be our last. We persist still because there is a deep and abiding truth about what we do and what we chase. In every bottle should be the best we have to offer. In every bottle should be the purest of thought and energy. It is our promise to those like us who understand the glory that is to be experienced from pressing ever deeper into our esthetic realm where pretense and untruth are fined away, where one is left, at the end, with an unadorned, unshining, but adamantine, unstainable monument to beauty, that the moment of creation is a shared one. That the making of the thing can never be truly complete is incomplete without the perceiving of the thing. I like this equation because it ties me closely to my tribe. My selfish desire... my self-sustaining need... to care for my tribe requires that the tribe (at the very least) acknowledge that there is an offering being made. The tribe will determine the quality of the offering, the extent to which it will take that offering to heart, will take it inside, will understand in its bones the seriousness and love with which it has been offered and will consequently judge its worth. No amount of outside influence will change the tribe's deliberation. All that can be done is for the one that sends the offering to send it unclothed and without condition, for the benefit of the tribe (if judged worthy) and for, ultimately, the purity of the giving itself. And this is the finest reward for the creator. It is not for the fanfare of the unknowing or even the approbation of the believing. It is for the bone-deep knowing that without him that thing of beauty would not be and could not be laid at Joy's altar.

Though I am not big on the idea of sacrifice, in part because I usually associate it with an unconscious and uncaring demand from those who are unworthy of receiving the gift in the first place, I accept that I am bargaining with my potential audience out of a position of weakness. What I have to offer (in the language of the trader and the language of the emotionally uninitiated) is 750ml of liquid that any one of the thousand other producers of wine can offer. This is strictly true. There are thousands of producers of wine in the world, many of whom can offer compelling wines at a fair price. My goal as a creator is to show as many people as possible that there is as wide a chasm between sublime and great as there is between great and acceptable. I am not here to pander to fashion or to provide what ten thousand others can. I am here to offer the very broth of Beauty, to ruin one for the ordinary, and to force an esthetic contemplation that will change lives.

iv

My father never lost his passion for truly great wine, even given the disappointed diaspora that was his and his brother-in-law's leaving the family business, Mirassou Vineyards, in 1984. With multiple generations of family involved, somewhat incestuously, in the operation, my father was unable to affect the quality of the wines being produced in any meaningful way. By this time, he had a large and excellent wine collection, and he was unable to look on mediocre quality without emotion, especially if he was tasked with selling it. Growing up I remember the rare occasions my sister and I would stay at his house, and after dinner I would watch him stand at the kitchen sink and pour small amounts of wine from different bottles into a glass and taste the blend every now and again. I asked him once, much later when I knew what it meant to make wine, why he didn't pursue the production side of the business. He told me he was more comfortable with people and liked to sell things.

In 1966 my dad and his cousins created a sales company to market and sell the family brand nationwide. Back in the '60s, there was only a handful of premium wine producers, and they were all trying to sell into a marketplace dominated by spirits and beer. As a comparison, 191 million gallons of wine were produced in 1966, 966 million in 2018—five times as much. Though Robert Mondavi hadn't yet achieved the worldwide acclaim he would in the 1980s and 1990s, those who knew him, knew, that he was a marketing and sales force of nature. This group of San Jose 20-somethings hired Mondavi to be their consultant. Every other week, they would meet in San Jose, and on one such occasion, in order to show the group how great wine could be, Robert brought the classic 1959 vintage of first-growth Bordeaux wines for the group to try. The Bordelais classified their Left Bank, specifically the Medoc, wines back in 1855, creating five separate *Growths*. The First Growth now contains five chateaux, the Second Growth more, and down through the Fifth growth, which contains the most and is the least prestigious because of it. These wines have long been the standard against which every other Cabernet-based wine in the world is judged, and it is a vinous and fortuitous rite of passage for young winemakers to get their fill of Château Latour, Château Mouton Rothschild, Château Haut Brion and the rest, as they are learning their craft.

Truth be told, being in the wine business, selling wine, sometimes even making it, doesn't necessarily give you any special insight into what it's all about. My family, except for my father, viewed wine as a commodity, something the family made, partly an inertial activity that was generations old and not inherently special. We didn't produce widgets, exactly, but wine production was what we did, not what we were. I don't know, firsthand, how any of the others felt about that day and those great wines. My dad has told me, at various times, that tasting those wines was a religious experience, and that it was like unknowing everything you thought you knew about fine wine and then struggling to envision a whole new textural vocabulary. This experience was so profound, in fact, that it was the first indication that he was on a different quality page than his colleagues and would eventually lead to my father's departure from the family brand and inspire the creation of The Steven Kent Winery with his son almost exactly thirty years later. It is my view that that momentous day in 1966 was the linchpin event that showed the potential of wine to inspire and that thrust my branch of the family into an aspirational wine current that goes back as far as the desire to create greatness goes—the very beginning.

My dad has had advantageous wine timing his whole life, and lived in the time when a normal guy, but passionate about wine, had the occasion and could afford to try the '47 Cheval Blanc, some say the best wine ever made. Imagine that, millennia of winemaking, writing about wine since the age of Pliny, and many smart and worldly drinkers can name *and* agree on a wine that might be the best ever, and you get a chance to try that singular vintage. More than once! The '45 Mouton (his favorite), gullets full of Domaine de la Romanée-Conti, the Inglenook wines from the '40s, my father has experienced them all. I don't have the capacity to be jealous; envy is another thing. My father had the opportunity to taste the most heralded wines of his time; the fact that he took advantage of it is a testament to his intelligence and his passion for amazing wines.

I've had the good luck to grow up in a family that farmed grapes and crushed them and fermented them, and cleaned fermenters out, and hauled pomace (the skins and seeds left over after pressing) to the vineyards for the cows to eat and bottled them and fork-lifted pallets to the warehouse and then went out to Chicago and New York and Cleveland and Jacksonville to sell them. Wine production was a practical exercise for previous generations of the family; it was what we knew how to do.

One of the great things my grandfather's generation did was make the next generation find their own way into the world of wine. The grandfathers owned the winery in the eastern foothills of San Jose and one of the vineyards we planted in the early 60s in Monterey County, and they told their kids that they needed to differentiate themselves from previous generations and strike out to new lands. They did and were successful for a while, until the clusterfuck that was the family dynamic rotted away at the piers and the piles until it all came quietly melting down.

It is pretty to think that my family was granted a peek at the future as it looked to acquire farming property in a more hospitable clime. Unfortunately for my generation, though, it was granted only that one peek. If they could have seen more or more clearly, they would have taken advantage of the opportunity to purchase 1000 acres on the Napa Valley floor. Instead, it is said, that one of generation number four told Bob Mondavi that we were going to find a much better place than Napa to put down these new roots. Soledad became the new home. Soledad, on the windswept Salinas Valley floor, home to row croppers, potatoes for the potato chip company and sugar beets for the sugar company, and a maximum-security prison too. An inherent, hubristic flaw in the family patriarchy, this was; a thousand acres in Monterey when there was no wine industry there. With the exorbitant valuations for land in Napa, there was no way to argue—financially—that Monterey County was a better choice than Napa, though, had the right grapes been planted there in the beginning, my family may have become known for the incredible, world-class Chardonnay and Pinot Noir coming from this area instead of for producing Cabernet Sauvignon that never got ripe (a writer for a major wine publication wrote about the "bell pepper problem" at Mirassou, referring to the chemical compound in bell peppers and Cabernet Sauvignon that gives the former its flavor and that typifies an unripe version of the latter if it doesn't get enough direct sunlight). Getting Cabernet ripe in this part of Monterey is extremely difficult, and that set the brand even further back.

The irony of the Mirassou and Monterey County story is that the area did become extremely successful, just not, long-term, for the family brand. It can be argued though, that if not for the vineyard research the family did with UC Davis in 1958–1960, if not for the Mirassou family taking the plunge and planting 1000 acres in Monterey at the beginning of the decade, the County that now has 40,000 acres of grapes planted and whose crop is worth more than $200,000,000 per harvest, the County that is home to names like Pisoni and Franscioni and Kendall-Jackson and *Mer Soleil*, would still and only be a row-crop paradise.

Only the truly great are granted the clarity that sees the present and near future as if through a glass pane. They just seem to know what the future holds or are happy enough to know that the future is out there uncoalesced until they gaze upon it. The family ended up making some pretty nice wine from these vineyards over time, but the battle they were fighting was never about the grapes from that early point on. The glass pane was opaque; the joy wasn't of sufficient quantity to guide them over the bad decisions and the internecine un-coordination, and the brand came to nothing with their stewardship. This part of the family's history in wine serves as much as a warning as it does a recitation of past events. In order for the family's business to move forward prosperously, all of the players need to be reciting lines from the same play; there must be general agreement as to the underlying direction—the goals, the overall sense of quality, the hospitality mission. There must also be a true love for what we are doing, for the business is too difficult without the passion to salve the everyday hurts.

············ v ············

My father had planned to take some time off after he sold his interest in the family business to his cousins, but an old friend who had once worked for him, named Ivan Fuezy, wanted to start his own brand. Ivan was a short, hirsute Hungarian with a personality much larger than his person, and a penchant for dreaming. He had escaped his native country as the tanks of the invading U.S.S.R rolled into Hungary in 1956. He made his way eventually through Western Europe to Ohio, and from Ohio to California. Ivan worked his way up through the industry as a salesman and as a wine broker. My father knew a lot of people on the winery and distribution side of the business, so he and Ivan created the brand, *Ivan Támas Wines* (using Ivan's first and middle names). Ivan Támas started as a négociant brand that specialized in value-oriented, *fighting varietals*—Chardonnay, Cabernet Sauvignon, and Sauvignon Blanc. Differentiating the brand with

California-Italian wines was a smart move as there was a rebirth of interest then in the varieties that the first wave of immigrant winemakers had brought with them from their homes all across the old-world peninsula more than a 100 years earlier. The Ivan Támas brand had a brief spurt of recognition for its Pinot Grigio and Trebbiano (under its French name, *St. Emilion*, the base fruit for Cognac) and was one of the brands trying to create a wave of interest in domestic Sangiovese, the primary grape of Chianti. One side of our property in Livermore was planted to Italian varieties; the other side planted, because of its utility and ubiquity, to Cabernet Sauvignon.

All through 1995, as I was contemplating coming back into the wine life, my father and I talked about a small, Cabernet-based wine brand whose mission it would be to make a world-class wine from the Livermore Valley. We both knew of the inherent quality of the Livermore Valley growing region, me through books, and him, through conversations with Phil Wente, who was in charge of viticulture for his family brand, Wente Family Estates, and probably the foremost expert, at that time, on the quality of the Livermore Valley appellation.

I began selling Ivan Támas wines in early 1996, having finally committed to working for my father and his partner. At the same time, my father and I continued our conversation about a small, high-end Cabernet brand. Working within the Wente wine facility at that time, we could taste a lot of small-lot wines. In 1997, my dad found a lot of Cabernet Sauvignon that was destined for inclusion in a much bigger and much less good offering of wine. We tasted this barrel group over and over again, finally picking the barrels we thought worked best together. With this wine from the 1996 vintage, The Steven Kent Winery brand was born.

Because of the non-compete clause in the sale agreement between my father and his cousins, we could not use *Mirassou* as part of our new brand. Consequently, *Steven Kent* was named for our first and middle names (we figured that it was easier to spell and say than *Mirassou*, which I knew could be mispronounced in a dozen different ways and because the perception of quality of the family brand had been headed downward).

Our first wine was released in 1999 and had the great fortune to come upon the market in one of those periodic ejaculations of unreason and profligacy. If the wine was good and the storytelling was at all persuasive, there was practically no way to fail. I know this in hindsight. Then, I was so far into it that I could only see the next buyer, the next chance to vindicate the vision.

I remember my very first sales call. The wine had gotten a great review in one of the magazines, and a local, very well-respected wine shop called to inquire about bringing the wine in. I blanch at my naiveté twenty-five years later. I had created this allocation scheme where only 2 six-packs would be given to any one account (I thought that I was one of these anointed Napa brands—what a putz!). With a 600-case production in the first year, absolutely huge for a brand like ours, and a price point double any other wine from Livermore Valley at that time, here I am restricting sales to 12 bottles (*geez*). Well, fortune and cowardice conspired to bring my head level. I brought the wine to the wine shop manager, and he loved it. I told him that I could sell him 12 bottles, and he asked if it was stupid to request 144 instead. Luckily enough, I was able to justify my capitulation with "this is a great start, I can tell the next guy only 12 bottles." So, much more often than not these happy accidents of fortune don't ever come along, and even if they do, they often amount to zero as far as the development of a brand might be concerned. They do stoke the imagination, however, of every salesperson worth her salt. Ivan sold his interest in the brand to me in 1997, and the success of Steven Kent in the early 2000s allowed me to shift nearly all my focus to this brand. Ultimately, in 2004 my father and I sold our interest in the Ivan Támas brand to the Wente family.

The Steven Kent brand lifted the reputation of the Livermore Valley from the beginning, in part because the quality of the wine was high and because I didn't think that there was any valid reason we should sacrifice on price even though we weren't a Napa brand. The price of our first bottle of wine was $45, which caused many of our compatriots in the Livermore to wonder if we were crazy or arrogant. When you are as new and small as we were, we had the luxury of being honest and self-aware. My dad and I spent a lot of time and considerable money buying Napa wines that we tasted blind with ours before we came out with the founding price. In practically every tasting, we felt our wine showed that it belonged in the same competitive set. We concluded that we would do no favors to ourselves in the long run if we came out of the chute apologizing for our wine by presenting it at a price that was governed by the current perceived quality of the appellation. I found over the years that no one thanks you for giving them "value," and the price is one of the more obvious ways of declaring your respect and admiration for the wine you are bringing to market (the weight of the bottle, as we would see a few years later with the release of the first vintage of Lineage | Livermore Valley in 2007, our Bordeaux blend brand, is also an important chapter in the wine-quality story).

Soon after we started, our volumes got to be bigger than what we could organically sell. Consequently, pricing started to become more flexible than it had in the past. I was on a plane jetting down the runway to Denver from San Jose on a bright Tuesday morning to work with our distributor sales team when the Twin Towers in Manhattan were struck and fell. Our plane came to an abrupt stop and taxied back to the runway and I watched, hollowed-out, the aftermath alone on TV back at my house. I didn't know quite yet, as I hugged my kids and wife after school, that with the destruction of all the lives and the buildings in New York that day that the fine wine market would existentially change for small family-owned brands, such as Steven Kent, and that I would need to find a more permanent and personal way to connect to an audience that would champion an upstart brand over one that could be found on every store shelf.

Sara, my youngest child, was eight or nine when she brought home a school art project that she called *Positive/Negative*. The school project was meant to teach the kids about the concept of positive and negative spaces. Using different colored construction paper, Sara created the reverse and inverse of various shapes and colors. In the process of creating a set of wine offerings we could sell to our customers in the tasting room, I started a wine club that would feature wines named for the artwork that adorned their labels. I wanted to repurpose the idea that Château Mouton Rothschild used to such wonderful effect: a famous artist would create a work specifically for the label of their greatest wine each year, making it one-of-a-kind. Our version of this concept would make each of our wine releases extremely intimate and personal as we celebrated the artwork of our family and friends.

The first wine we released was *Positive/Negative*, from the first Cabernet Sauvignon produced from our Home Ranch vineyard. This wine led to our first Sangiovese called *Cherry Blossoms*, named for another school art project made by Sara's older sister, Katherine, featuring tissue paper blossoms glued to branches made from the dripping of black ink on white paper. Our wine clubs continue to save our financial ass while providing really delicious wines exclusively for our members. The advent of our first wine club tilted the relationship we had with our customer. In practically every circumstance outside of the chance encounter with someone stopping by our new tasting room, we did not know who was buying our wine. The wines would leave, in anonymous hands, a store shelf in Del Ray Beach, Florida, and I would never know the name attached to the "depletion" that bottle represented to my distributor. As more wine sold through the tasting room and we created relationships as well as took in more money because of the retail vs. wholesale price structures, the greater the value the wine club contributed to the company, and the more we began to change our sales priorities. ○

CHAPTER
5

MAKING WINE IS MORNING WORK

Making wine is morning work. The heart of the winery is always beating well before the sun rises over the eastern hills as trucks carrying fruit from all over California arrive at the scale house. The fruit that was picked in the perfect cold blackness a few hours before is arriving now on a flatbed truck. Headlights dance in the cold air as tank trucks and pickups pulling trailers and trucks hauling twenty-ton gondolas samba slowly to the scale. The mechanical clatter of the crusher bats off the high masonry walls of the winery as Antonio and Joel, two of Wente's small-lot winery crew, get the fruit-receival machines up and running. I head back to my workspace in the tasting lab with the weigh tags from the morning's load. Aidan is off-loading half-ton bins of fruit to the scales and Beth is readying the fermenting boxes. In harvest time at the crushpad, an unruly ballet is always in progress: forklifts zoom like hummingbirds taking picking bins off trucks, those trucks heave out of the yard back to the vineyards to be loaded again, forklifts lurch through large swinging doors bringing full fermenters to the 100 room (the Wente Winery is divided up into separate work spaces and each is simply and clearly named), the winemaking team is racing up tank ladders carrying buckets of yeast redolent of the morning bakery, the foreman is yelling in Spanish and his words form curlicues in the cold mist of the morning. Beneficent chaos. And noise.

I blow on my hands and stop by our bins of fruit stacked like bars on a graph to look at the bunches and to taste individual berries. The fruit today is perfect. Small spiders meander over the shoulders and wings of bunches, and I flick them away as I throw a grape into my mouth. The fruit is cold. The juice is sweet but not too sweet. The skins are firm but not too firm. I spit the seeds into my hand and look at the color: mahogany. Perfect. If the pickers have done a good job in the middle of the black cold night, there will be few leaves in the bins. With the new equipment most of these will get blown away by little puffs of air and shunted to a bucket where the garbage goes.

In the lab I grab another cup of coffee and fill out my harvest book with the crucial information about each delivered load of fruit. Growers have to be paid, and the weigh-tag information I transcribe in my book will ensure the correct dun. I enter the vineyard name and block number, date and grape variety, the number of tons, and do the simple math that tells me how many gallons of juice I should have. If we receive more fruit than we expect, we add another fermenter to the line sitting at the end of the crushpad. Or we'll take one away.

My phone is handy so my team can tell me when our fruit is about to be processed. One of my jobs each day is to walk through the growing garden of fermenters, lifting lids and smelling each to make sure that there are no "off" aromas that tell me there are problems with yeast. If you crush a vitamin and put a little of the dust on the tip of your tongue, that bloody, feral flavor is what troubled yeast smell like. Most of our fermenters are plastic boxes, and they hold about 1.5 tons of fruit. Deep into the season, we may have as many as 20 fermenters in varying states of *done-ness* lined up in the 100 room. The cellar in the winery is always about 55°, which is warm compared to the temperature today at the crushpad outside. The 100 room is a concrete room with concrete floors that slope down from the middle to drains on both sides. It's a hundred years old and where we do our fermentations. I take off my sweatshirt because it's warm enough to, and I lift the lid off a Cabernet Franc box and feel the heat coming off the mounded fruit. I wave my hand from box to nose a couple of times and sniff. I smell the doughy notes of yeast and the sweetness of fresh fruit. What I don't get—because I've gotten it too many times before—is a scalding lungful of carbon dioxide that shows fermentation is well under way. That's why the hand-waving. I go up and down the rows, smelling each box, looking for colonies of mold that might have formed overnight and choosing which boxes are ready to be punched down. The mold used to scare me, but I was young then. The first punch down gets it.

One of my other jobs in the cellar is to prepare yeast to get fermentations started. If you've made bread from scratch, the process will be familiar. I choose a yeast strain that works well with the variety in the fermenter and rehydrate the freeze-dried granules with 105° water. After twenty minutes, I add some of the grape juice from that fermenter to the bucket to feed the yeast and to lower the temperature of the solution. The fermenters are cold, and the difference in temperatures between the bucket of yeast and the fermenter full of *must* (unfermented grape juice and skins and seeds) needs to be minimized in order to give the yeast the best chance at a strong and problem-free start to fermentation. I add juice every 10 minutes for about 40 minutes; the vigorous bubbling of the bucket and the warm, doughy aromas are the proof that the yeast are multiplying and happy. I spread the yeast solution over the top of the cap of skins when the temperatures are right and throw a lid back over the box.

We don't punch down the fermenters right away because we want to get the flavor and aromatics contributions from the yeast and bacteria that are stuck to the grapes when they are harvested. These strains start fermentation, then die when the solution reaches about two percent alcohol. This technique allows us to capture the ethereal organoleptic endowment of these "wild yeast" which help us build the complexity we are looking for. When the cap of skins begins to dry out and crack after four or five days, we start the punch-down regime.

Once the yeast culture hits the juice it is like a needle of adrenaline to the heart. There is so much sugar there in solution to serve as food for the yeast, typically nearly 25% sugar upon which our strain of *Saccharomyces cerevisiae* will dine, that it will engorge itself, budding and re-budding until the number of uni-cellular inhabitants in our fermenter grows dramatically. The yeast's growth translates itself from the realm of the unseeable to the clearly obvious across nearly all the senses. In barrel-fermented Chardonnay the hiss of carbon dioxide escaping through the fermentation bung sings a tune of yeasty imperiousness while that same CO_2 in an open-topped bin hits you like a caustic physical wave. D254, a strain of yeast we use often in Cabernet Sauvignon fermentation, gives off a chocolatey note that is nearly identical to that of the *Yoohoo* I drank in college. And walking through winery doors into a room full of fermenting wine is to experience the aromatic equivalent of moving from silence into the practice room of a full-blown orchestra.

The number of different scents is nearly uncountable, and you can know a lot about the quality of a vintage just by letting that wonderful stink of fermentation wash over you. If things are progressing nicely, the yeasty transformation of fresh fruit into much more complex wine will dominate. The robustness of Cabernet Sauvignon fermenting plays a bass note, powerful wine inchoate, in notes of dark and black fruit. Sangiovese smells redder, the fruit more fragile and tenuous, an earthiness entangling itself with that fruit in an invisible helix of *esters* wafting ceiling-ward.

Each time I walk into my own fermentation room early in the morning, I am instantly transported to my childhood days working at the family winery, running through rooms filled with old redwood upright tanks aging wine and filling those rooms with a wonderful and somewhat foreboding perfume. Those rooms were full of mystery and full of something profound and delicious, even to an inexperienced 12-year-old. This connection to my youth seems ever more instructive and profound with each vintage. It is almost as if all that one would ever know or learn was already encapsulated in the body and mind of a child, and the passing of years and the doing of things one attributes to experience is really just a folded-back reflection of those times of innocence when the world was full and the joy, perfect.

ii

The bins have been moved from the cold room (a room where tanks of wine are lowered to below freezing to precipitate out excess tartaric acid and thereby increase stability (nobody likes to see a bottle of white wine in her refrigerator with garbage resembling parmesan cheese submerged in it) to a warmer room (about 55°) where the wine yeasts then start getting active: multiplying their numbers and producing heat as a by-product of their activity. Fermentation follows a curve over the week or so it takes to get a bin of red wine dry. When fermentation is at its most active, you can see sugar drops of 7–8° Brix per day at first, then tapering off as the amount of sugar is diminished. All told, it takes about a week or slightly longer to get red wine dry (the condition in which wine has no or very negligible amounts of sugar left). Over the course of this vinous week, the juice is obviously less sweet but there are numerous additive qualities given over to the wine that would not exist without the destruction of the grape's mojo and internal architecture. The wine grape is one of the sweetest fruits in the world and is ideal for producing THE greatest beverage in the world.

Fermentation is a magical process in which countless chemical reactions are occurring for the wine maker's purpose of getting sugar to alcohol. In this chain of events, some pretty smelly things can happen. Our unicellular heroes—yeast—responsible for this transfiguration of the pedestrian to the sublime require certain nutrients to work at peak efficiency. If they don't get the nitrogen they need, their malodorous protest makes it known. Aside from the issue of stink (which is easily remedied by the addition of diammonium phosphate [DAP]), the more serious side-effect of yeast malnutrition is the potential for ferments to get stuck—for a wine to get only partially dry (there is nothing worse, or less saleable, than a sweet Cabernet Sauvignon). The other common issue with early ferments, especially when the juice and skins have yet to truly begin the sugar conversion, is the presence of volatile acidity or *VA*. All wines have some volatile acidity, and it can take the aromatic forms of vinegar and fingernail polish remover (ethyl acetate). At its worst, too much VA turns a wine, through bacteria that are ubiquitous in wineries and often spread from one fermenter to another on the feet of fruit flies, into vinegar. At lower concentrations (there are legal limits to VA, but the derivation of that limit seems very arbitrary when taken outside a more inclusive chemical context), volatile acidity can lift a ponderous wine up a bit, can leaven overripe fruit, can make an older wine taste younger. The legendary 1947 *Cheval Blanc*, thought by many to be the finest wine in history, was illegally high in VA. That precursor to vinegar made a very rich, viscous wine harvested from a very ripe, hot year a little less so.

All this said, the best tools the winemaker has are the personal ones. Each morning we go through each fermenting bin, lifting the lid, and smelling for any aromas that aren't what they are supposed to be. Hopefully, very early on in the process, you have the scents of fresh fruit and yeast. And we'll also taste. In low-fill bins, this requires a small step ladder (*¿Donde esta la esclarita?* is how you find one). You want the juice/wine, so you dig through the cap of skins, which is usually about 12 inches deep in this kind of vessel, once fermentation really gets going, until you reach the liquid. Then you splay your fingers over the top of the beaker, leaving only the smallest margin between the digits so you get as much juice and as little of the solid stuff as possible. Armed with our senses and a notebook, we write down what our nose and mouth are telling us about sweetness, about fruit, about levels of tannin, about the heft and duration and complexity of the mid-palate of the wine, about the balance of acidity to fruit to tannin on the finish. These elements, these Knights of the Wine Table, are, at once, what the winemaker is trying to protect and what will allow the best of wines to protect good taste for years to come.

In all the screenplay writing books, they tell you that there needs to be tension, that the protagonist has to change over the course of the movie in order to be believable. On page 20, the [A thing] happens and on page 120, the [Z thing] happens, bringing a conclusion to all that came before it. Harvest allows us the opportunity to tell multiple stories, filled with all the heart-wrenching tension and evolution of story that each separate fermenter contributes to the book of the season. With luck and perseverance, this book will be one of many volumes.

The color of Chardonnay and Cabernet juice is the same. Cabernet is red because we ferment the juice with the skins. It is the skins that give over the color and tannin and flavor and structure. If you think of a Cabernet fermenter as a cup of tea, you'll see how the repeated punching down of the grape skins under the surface of the fermenting must, like the dunking of the tea bag, will increase the color and tannic content of the liquid. In the plastic fermenters there are no valves, so we have to take a seven-foot-long metal pole, stand on the edge of the bubbling box four feet above the concrete, and punch down the cap of skins into the wine below.

Punching down is one way to get at the color and structural elements of the fruit. Pumping over is another. By attaching a hose to a bottom of a fermentation tank, you can pump that wine and spray it over the top of the cap. The wine, on its way back down to the bottom of the tank, will grab on to anthocyanin and other molecules leeching from the skins of the grapes and infuse them into it. During the week or so of primary fermentation, when yeast convert sugar to alcohol and carbon dioxide, we punch down or pump over three times a day. With lighter red wines, the routine will vary, but with Cabernet Sauvignon and Petit Verdot (another hearty Bordeaux variety), for instance, we punch down or pump over every six hours or so.

The wine changes each day, but it is during extended maceration, after primary fermentation is complete, that the daily changes are ravishing. *Extended maceration* is a technique we sometimes use to build structure in our wines. Normally, we will press a wine off the skins soon after it is dry and pump it into barrels. On occasion, though, after tasting the evolving wine in the fermenter day after day, I believe the wine would be more complete and ageworthy if it had more tannin. By keeping the wine in contact with the skins and seeds for an extended period of time, the alcohol, which acts as a solvent, breaks down the physical material it is in contact with, leeching out more tannin over time. We continue to punch down every day, even when the cap has dropped under the surface of the wine, leaving only small atolls of seeds floating lonely on the purple-black expanse.

Extended maceration is all about *feel*, so there is no codified endpoint. It is the un-linear quality of day-to-day progression that is the most fascinating thing about this process. We get several days in a row where everything is moving beatifically forward. Tannins are broadening and becoming more defined, the fruit is moving from black cherry to cassis, and there is a general sense of fullness and ripeness to the wine. And the next day, it all goes in the shitter. The tannins are harsh, the finish is short, there is NO fruit—this wine is a disaster! What the hell are we going to do with this crap, it's 200 cases of very mediocre Cabernet!? There is a next day, however. And, usually, the beast is beautiful again.

It would be reasonable to conclude that bins containing the same grapes but inoculated with different yeasts would begin primary fermentation at different times and would reach that golden moment of NOW and be pressed on different days. You'd think, though, that the same grape, from the same vineyard, inoculated with the same yeast would progress in lockstep with its bin-mate. More often than not, that is not the case. The minute differences in nitrogen levels from one fermenter to another, the slight temperature differences in different sections of the 100 room, the state of fermentation of neighboring bins (more mature fermenters throw off yeast, especially during punch downs, like a virus shedding itself) will all affect the kinetic curve and timing of primary fermentation.

I have been making wine since 1996 and didn't know anything back then. What I know now, I've gotten from reading a lot, and drinking a lot, doing a lot, and taking lots of notes. I only use a TOPS Computation notebook with page numbers that is bound with a spiral so I can fold it over. This morning, as I do every day with every box, I write about how the fermenters are behaving. The notebook gets filled up over harvest and gets more and more stained. Past books look like murder scenes. I have a beaker and a wire mesh strainer that fits over it. I dig through the warm cap of skins and seeds that float on top of the wine until I hit liquid. I taste the wine and write down what I taste and how the wine feels. The feeling part is the most important. A typical entry reads like this:

> LR-34. CAF. D254. Bit redder than yesterday. Herbal notes moving from fresh to dry, more persistent//dried chili flavors today, red fruit. Acid better today. More richness…acid in mp to finish more defined. Great length. Press in 2–3 days?

The way the wine tastes and smells, its organoleptic qualities, evolve more predictably and more slowly than its textural properties. I make pressing decisions based on the feel because I don't want the barrel, that will make significant structural contributions, to overwhelm the way the fruit is expressed in the wine. If I get most of the way to the structure the wine demands in the fermenter, the barrel will play its proper role as a relatively neutral scaffold that supports rich fruit.

And the wonderful (and maddening) thing about this exercise (much like that of the making of wine itself) is that it evolves every day; with very young wines still in the fermenter, devolution often seems much more accurate a description. There is only one objective reality about wine: there is no point at which the wine is unmoving. What we are looking to accomplish here is to find that point at which there is harmony. We want richness of fruit and tannin; we want length and roundness; we want black fruit in balance with structure; we want those elements that will allow the wine to age with grace, but not too quickly. We note all of these things against an internal compass. With our finest Cabernets, we understand that this is only the first (and shortest) step on the journey (once we decide to press the wine off the skins, we cannot unsqueeze). The time aging in barrel is going to change this fledgling wine immensely, so we factor this into the decision about when to pull the plug on extended maceration.

The fermenting bin sits in a plywood frame so that we can get 1.5 tons of fruit into without it collapsing. The bins are made by a company in Fresno, and they cost about $350 each. Each fermenter has a plastic top that we can use to help regulate heat; we keep it on when the fermentation is on the cool side and take it off when the must rises to a temperature (as a physics problem dealing with the mass and heat transfer) that might harm the yeast. The bins are tall enough that I have to stand on a small step ladder to adequately punch the contents down each day. My younger crew can stand on the edges of the bin, like circus performers, the thin edge of wood and plastic taking the place of the wire strung between poles. They get more leverage, and punch down more efficiently than I. The fermenting bin is its own world; it is born, and it dies on its own scale of time. The bin is where the transformations happen, where the steps from potential decay to apotheosis occur. All of the interim aromas and flavors that separate beautiful wine from a rank container of volatile acidity occur in this small (though, philosophically—an infinitely large) box. These boxes, holding the work of the year, are our most intimate and valued possessions. As our most fundamental mission is to take care of people and to relieve them of even just a small amount of daily strife, the fermenter box is where the vehicle of that hospitality mission (our magic elixir) comes to, alchemically, *be*.

As with every other facet of wine, each fermenter is always in the process of articulating its own moment of apotheosis. It is the winemaker's job to watch for that moment of becoming and to help it to fulfill that destiny, to be only—and exactly—what it is meant to be. Unlike the beer makers and the spirits producers, we only get one shot a year to make wine, consequently there is a ton riding on making sure things are done correctly. The higher the price of fruit, the more the ass puckers at any sign of malodor or sluggishness in the fermenter. The older you get, and the more boxes of fruit you ferment, the more you understand that there is, running beneath the veil of certainty that science is supposed to convey, an atavistic sense of *time* that is beyond measure but that is perfectly in sync with the needs of the must bubbling away in the bin. The secret to sanity (and quality), then, is to learn to let go of what cannot be counted and to trust in your benign neglect AND your watchfulness. It gets easier over time to let that clock, the one closely tied to all the preternatural things at the beating center of the earth, wind down according to its own wonderfully un-modern, anachronistic and perfect pace.

Aidan calls to let me know that our fruit is next in line for crushing. I put my sweatshirt back on and leave the quiet and warmth of the fermentation house for the cacophony and cold of the crushpad. Beth and Aidan are standing on either side of a conveyor belt that brings bunches of fruit from a metal hopper to the crusher-destemmer as I walk through the massive wooden doors to the crushpad. They are removing unripe fruit and leaves and broken pieces of vine, insects, the occasional lizard, from the fast-moving carpet of grapes in front of them and throwing the garbage into plastic cans by their feet. Beth is blond and tiny and a dancer and strong. Her arms and legs are thin but all that's there is muscle. She is smiling as she normally does, and her hands are moving fast. This is Beth's first year with us. She grew up in Oregon but lives in Livermore now with her teacher husband, Patrick. An acquaintance was taking classes with her in the viticulture program at the community college in town and knew she was looking for a harvest job and that we needed the help. Aidan has headphones on, and he is moving a little to the silent music. I dodge the hoses and cords that litter the crushpad and climb up the ladder to start sorting fruit.

The grapes come in waves, larger clumps as a new picking box is first dumped into the hopper. I am focusing on flecks of green in an otherwise crimson carpet; this unripe fruit won't help to make great wine, so it has to be thrown away. The fruit is moving fast. It takes a while to get used to the speed and to the stickiness of the sweet fruit, but you do. I take the emotion out and just let my hands move, clutching at as many leaves and green berries and pincher bugs as I can. I try to keep my vision in a small frame because moving your head too much can make you feel like you're on a boat. Two hours after we start, the sun has breached the wall of the winery and shines down hot off the stainless-steel tanks and presses. The last knot of fruit goes by. I dip my hands in a bucket of water, dry them on a rag hanging from my back pocket, and drop down from the ladder. I steady myself, grab a slug of water and go off to the next thing. The fruit-receival part of the day ends when Aidan forklifts the last box into the 100 room. At night, hours later, when I take my last look around, the boxes are in rows with their lids on. The garden keeps growing. o

∞

Tasting from the Fermenter

∞

CHAPTER
6

REAL CABERNET FRANC

I remember the first time I tasted real Cabernet Franc. I had just finished Kermit Lynch's great wine book describing how he built his business, and I was compelled to taste the wines he had discovered in his travels. I bought some of the '09 and 2010 vintage of what he had and knew that I had found a vinous signpost for myself. The wines of Joguet and the Bretons, eminent producers from the Loire Valley, flaunted their austerity and shoved their acidity right up in your face, fruit parading like a hooker in the old Times Square. These producers made a number of different wines; all had these things in common: they were 100% Cabernet Franc, they had an undeniable energy and sexiness, and they were flat-out delicious. Like the best of wines, these bottles created an emotional connection with me that continues to thrill.

I don't know the producers personally, and don't know that I will ever meet them, but I imagine them to be kindred. If we sat down over a bottle of wine in a restaurant in Livermore or Paris, we would be talking about how we express the fruit we have most authentically, what it means to devote oneself to a specific variety, how that quest impacts the business we are trying to transact and if that quest leads to some kind of balanced life. I would lean on them for their expertise in this great and grand variety, so much more expressive than its child, Cabernet Sauvignon, and though I may half-heartedly argue that multiple varietal representations of the wine life can lead to a wider and more hospitable view of wine's magic, they'd counter that no other grape offers such an abundance of *soul* when it is done correctly. They would, of course, be right.

Unmask the grape, leave away the wood and the over-extraction; eschew bigness and choose instead just the simple, unadulterated, the *I-get-up-out-of-bed-looking-like-this* gorgeousness, and you have the truest expression of Cabernet Franc. No other grape is as consistently surprising and reaffirming at the same time. When you drink Cabernet Sauvignon consistently, especially from the same region, what you get are differences in amplitude, not differences in the strings and heart of the wine. Hey, singer X is great and all that, but there is nothing to be learned from him if all you do is change the volume. That's Cabernet Sauvignon. Cabernet Franc is that singer, but that singer if he changed to a she then changed back to a different he, was French, then sang the ululations of certain South African tribes, then sang the operas of Puccini, then yodeled as none has ever yodeled before.

The best Cab Franc continually resonates at a pitch and frequency just out of the range of comfortable explication. It is the great conniver, is Cabernet Franc. At one moment it is as prosaic in the nose as mediocre Merlot—all cherry and wood. In the next though, it's as mysterious as the first time you got her panties off. It smells of ash from a day-old fire on the beach or bitter chocolate ladled over with brandied raisins. In the mouth, the best of wines make you feel as if you're on skates being blown across ice—the blades cut deep but all you sense is the thrilling movement forward; that's Cabernet Franc on its second-best day. Cabernet Franc could be all corset and hair-bun if she weren't such a sexy bitch. Just one mouthful of really good Cabernet Franc and you know she knows a hell of a lot more about all things sweaty and carnal than you ever will. She enters slowly and dances languidly at first. You turn your head away, disinterested. It's then that she preens and polishes; her movements, then, become precise and rhythmic and self-possessed. As the story unfolds in the inflection of hip and the accent of breast, you are compelled to look at her true and are… lost. Pinot Noir is all intellect. Cabernet Franc is about orgasm.

ii

The *Judgment of Paris* taught us that there was little difference in quality and characteristic between Cabernet and Chardonnay made in France and in California. Some of the best tasters in the world: restaurateurs, wine writers, domaine owners—experts all, changed the story of wine for the world, not just California, that July day. In 1976 to celebrate America's Bicentennial, Stephen Spurrier, an Englishman who had a wineshop in Paris, had heard about the rise in quality of New World wines from his Californian assistant. To promote his store, he got French and Californian wines together, and the winning wines warped time. Golden State wines would have eventually ascended the ladder of public visibility and critical opinion, but Spurrier's publicity stunt was like the meteor that killed the dinosaurs and ushered in the new age. The results also showed something more prosaic: two of the finest grapes grown, Cabernet Sauvignon and Chardonnay, made wines practically indistinguishable from each other despite 6,000 miles and a thousand years of separation.

While there were many factors at play in this legendary tasting that led to this conclusion, one of the most interesting to me is that it showed that Cabernet Sauvignon and Chardonnay have a fairly narrow range of flavors and structures (if one forgets and forgives outlandish winemaking interference) while others such as Sauvignon Blanc from New Zealand and Cabernet Franc from Chinon are about as idiosyncratic and singularly identifiable as grapes get. Had the group chosen to taste Pinot Noir blind or Cabernet Franc, Sauvignon Blanc, even, there would have been little doubt as to the provenance of wines they were tasting.

Cabernet Franc was found by UC Davis professors in 1997 to have come up on unsuspecting Sauvignon Blanc in some tangled thicket, and this vinous booty call birthed the noble Cabernet Sauvignon. It's a wonderful trick of Nature that a grape that is synonymous with the most substantial and age-worthy of red wines—rich and tannic, breathlessly undrinkable while young in its Bordeaux incarnation—is the offspring of a white grape and a grape that, in its purest state, is fulsomely red in fruit, spritely in acid, smelling of a day-old bonfire on a beach, with a precision of structure and flavor that just knocks you out. I started making Cabernet Franc in 2005, though I didn't really know what I was doing. Lynch's wines, an evolving palate, and the cold season of 2011 conspired to put me on the right track, however. This amazing grape has become so important to us, and to the future of the Livermore Valley, that we started a new brand called *L'Autre Côte*, that will comprise only Cabernet Franc from the most felicitous sites. Northern France is not California, of course. Differences in temperature, length of the growing season, clones of Cabernet Franc, age of vineyards, historicity of the varietal itself all separate these separate places. A spell of bad weather in 2011, though, helped me luck into a style of Cab Franc that reminded me a touch of those glorious Loire Valley wines that opened my mind half a decade earlier.

The wine business can be complicated and making great wine has much more to do with feel and philosophy than it does with science. There is one monstrously important factor, though, that we can rely on in California that gives us a huge advantage and that takes away a lot of the risk with which agriculture is inherently saddled: we have great weather. While winemakers in Burgundy are worrying their fingers to nubs wondering if the solitary cloud on the horizon will bloom black into a harvest—and livelihood-destroying hailstorm—the California winemaker many more times than not gently rouses himself from his sun-soaked afternoon nap, puts his finger in the air, and decides that it's probably time to bring some fruit in. An exaggeration, yes. But not when describing the string of years of clement weather we enjoy in the Livermore Valley. 2011 was a different beast all-together.

In the winter we hope for rain here. Livermore only gets about 14 inches of rain per year, and in the decade of the '10s there were a number of years where the totals were dramatically lower than this, and December through March were bone dry. The winter of 2011 started, though, with torrential rain, vineyards couldn't soak all the water in and much of it rode off away in the arroyos that bisect our vineyard. Along with all the water, the weather in April and May (during budbreak, bloom, and flowering) was unseasonably cold. The grapes got a late start, mildew and mold pressures grew in the vineyard with the days of rain and the lack of warm weather to dry the vines out, and, as the season progressed, signs of ripeness in our Cabernet Sauvignon blocks were difficult to discern.

Unripe Cabernet Sauvignon is an ugly thing. It can taste of bell peppers and stewed asparagus; Sauvignon Blanc, one of its progenitors, afflicted with similar concentrations of methoxypyrazine, can smell of jalapeno peppers and cat pee—undelicious. Pyrazines, as they are commonly called, are compounds that give bell peppers their characteristic aromas and flavors, and they also live in the skins of the Cabernet family of grapes. With direct sunlight, UV radiation metabolizes these compounds out of the skins and very ripe Cabernet loses all of this herbal quality, substituting uncomplicated draughts of jammy and ripe berry in its place. Similarly, Sauvignon Blanc changes from this funky pepper/urine-stained dreck to a citrus-addled, wonderfully clean and refreshing ravisher. This relationship between direct sunlight and pyrazine has an overly dramatic effect on Cabernet Franc. Compared to Cabernet Sauvignon, Cab Franc's body is lighter and more transparent. Unbalanced pyrazines can color Cabernet Franc like a bad tattoo, and as permanently. Because working the edges is where all the risk and all the fun is, I strive to pick Cab Franc just when it has crossed over from the dank side into the light. I want herbs. I want my Cabernet Franc to be infused with the unmistakable redolence of fresh bay and sage. The picking decision, then, is crucial. Too early means the green of rotting vegetables; too late is flaccid and uninspiring pie filling. Getting it right is the difference between the supple muscularity of youth and the flab of the athlete gone to seed.

Though it rained throughout much of the summer and made the harvest of several of our Cabernet Sauvignon blocks a muddy mess, the growing season was long enough that we were able to pick enough fruit at sufficient ripeness to make some of the best wines we have made. We weren't able to make a lot, as much of the fruit never recovered from the ungentle start to the season, but what did come in was gorgeous. Our 2011 Cabernet Franc thrived. Harvested in a short window between storms, the fruit came in several weeks later than the preceding and subsequent vintages and arrived at the winery disease-free and tasting beautifully.

In the cellar, fermentation proceeded without a hitch. As the wine approached dryness, I would shove my hand through the cap of skins and grab some of the still-fermenting wine with a beaker and taste its progress. Each day, the wine seemed to grow darker and richer. What would become notes of soy sauce seven or eight years after the harvest could be gleaned in the increasing substantiality of the wine. There never was any doubt, though, as to its varietal provenance. Those thrumming notes of bay and sage, beneficent and pure, leapt from the glass, and the wine danced energetically on the palate under the influence of graceful acidity. 2011 was a sex bomb! With age the wine showed more and more complexity of flavor and aroma while sticking beautifully to its austere structural core. Wow! One of my favorite wines.

Sometimes you get lucky in these kinds of conditions, and a winemaker's mettle is tested. These years are the most fun because you are riding such a narrow edge. Missing the pick by even a day can be catastrophic when the next weather front may be only a few hours away. In California, winemakers can add water and tartaric acid (among many other things… practically all of which we eschew) if the fruit comes in a little too ripe. It is much more difficult to ameliorate a fermenter-full of fruit that comes in significantly unripe. Though the marginal years lead to more challenge and more seat-of-your-pants winemaking, I think I'll stick to the eight years out of 10 where the year is so good that if you can't make great wine then you really should be cobbling shoes.

The intervening years from 2011 to today have been mostly fortuitous as the weather goes. The one exception for us was 2015. And here, again, we made a Cabernet Franc that skirted that ripeness line all throughout fermentation to come out on the other side, smoking hot! I took a calculated risk on the picking day hoping that the lower sugar levels on the bottom of the block would be balanced out by the slightly higher sugar in the middle of the block. I was hoping for 22° brix and ended up with 20. *Quelle catastrophe!* 20° brix is where you would pick Pinot Noir or Chardonnay or Barbera for sparkling wine, not Cabernet Franc for… well, Cabernet Franc. When we got the wine into barrels it was green as a Leprechaun's civvies. We decided to add just a bit more new oak to the blend as wood barrels have a way of integrating with pyrazine to reduce its influence. We stirred the barrels weekly as if we were making Chardonnay.

Mixing the lees frequently and letting this solid matter descend slowly back to the bottom of the barrel has a way of adding a touch of richness to the wine and mitigating the green as well. Each week I would come to this pyramid of barrels and taste, and—at the beginning—I'd be heartsick, not only trying to figure out what to do with this fuck-up, but also lamenting the loss of a season's worth of my favorite grape. Everything can be done *correctly*, and failure may still be the result. Sometimes, too, your intuition leads you down an errant, twisty path that drops you, eventually, to the spot you steered toward at the start.

Over the months however, what was bell pepper evolved into the clean and dried herbal notes that bespeak true Cabernet Franc. The texture of the wine also evolved from knife's-edge acid to something slightly less sharp, from shards of glass to freshly laundered cotton. This evolution did not proceed in a straight line, to be sure. There were weeks that the wine was barely drinkable and others when it sang like a young Nina Simone. Like all wines with potential, the 2015 Cab Franc *became* at its own pace and took a shape completely in harmony with the blocks that were there, in the beginning, to build. I count this wine as one of my favorites. And if quality comprises careful shepherding and proportionate freaking out and an acceptance on the part of the winemaker that what is there is beautiful, for all its idiosyncrasies, then this wine is, too, one of my successes.

It is an interesting twist to the global warming story that causes me to be so certain that the Livermore Valley will be a spectacular growing area for Cabernet Franc. Livermore is dependent upon and circumscribed by the wind that comes in daily from the San Francisco Bay. Due to the increases in temperature in the Central Valley to the east of Livermore, which serves as the engine that powers the vacuum that sucks in this cool air, wind enters the Valley earlier in the day, abbreviating the amount of ripening that happens that day. The vine protects itself from the moisture-robbing characteristic of the wind by shutting down its microscopic pores, called *stoma*. The stoma regulate water usage, but they also restrict the amount of CO_2 the plant can take in. Carbon dioxide is used in photosynthesis for the production of sugar. With reduced sugar production each day, it takes longer for fruit to get to optimal ripeness.

Earlier ripening varieties like Cabernet Franc (compared to Cabernet Sauvignon), will end up being privileged by these conditions.

It does not matter that the notion of a grape being associated with a specific place is a European one, the ability for Livermore to be associated with a world-class variety (like Napa with Cabernet Sauvignon) that it "owns" in the mind of the wine press and the consumer has business implications that could allow a small region like Livermore to thrive. Our L'Autre Côte brand will be dedicated, then, to making stunning and delicious Cabernet Franc and proving again the validity of the terroir-driven model. o

CHAPTER
7

BLEACHED AND SILENT AND SOLITARY

In 1910 Kate Bankowski, lately a widow, cut her house in two and hired a gang of vineyard workers to roll one part of it on eucalyptus logs 300 yards west on Tesla. She found her husband hanging from a fig tree by his neck the week before, so he couldn't stop her. The house had been done up in the Livermore Victorian style with a wide porch that wrapped around the whole of it, a moat against direct access to the interior. To the south, like the tail of a comet, 80 acres of vineyard extended the parcel all the way out to Mines Road. The vineyard was planted to a large number of grapes, as was the custom then, and the wine was made from all of them—the field blend of young California.

"Klondike Kate," as she was known, took up with a field hand who lived with her in that severed house. The townspeople didn't like it but attributed it to the fact she was born to the outskirts of town. They didn't like it when she ran young girls out of the back of the house either, but that didn't stop some of them from taking them up. Kate and the field hand left the property suddenly and it was sold in probate to someone from town. The first World War came, and part of the vineyard was sold off. The next war arrived, and the vineyard contracted again. Families lived in and left the property over the decades, and each time they did, the vineyard, like a drying sore, shrunk on itself. In 1997, there were no vines anymore around the house and the remnants of the field behind it were pulled out a couple of years later. The misshapen house, now much less than half its former splendor, hunkered down in dirt. The family that owned it last before us sold the property when the kids went off to school. From 80 acres of twenty different grapes planted on the finest soil in Livermore, the hurt house lay bleached and silent and solitary in the middle of eight purposeless acres. That was when we bought it.

Three-quarters of those acres were planted to vine right after we acquired them. There are a few buildings on the property, the largest comprises the old, severed house refashioned into an Italian villa. On the top floor are offices, meeting rooms, and a kitchen; below is a low-slung concrete box that is our tasting room. Connecting Tesla Road to the main building, 100 yards off the street, is a dirt road that forms a leaning circle. There is a promenade of olive trees that separate the in from the out, and on each side of the entry are vines. There are a few patches of asphalt buried under loose dirt on the exit side, but every year, as now, the potholes on the entry side bloom like inverted mushrooms. Tracey Ramirez, the late boyfriend of Jennifer Fazio, my VP of Operations who has been with me since the beginning, would fill them and grade the entry in the spring, but they come back. The winter rains turn them into cups.

The patio sits on the eastern side of the main building, timbers from the old house still framing the cellar roof. Looking south at what had been part of the original Wegener (the first owners of the property and the builders of the original house) estate, there is a 50-acre rent between the vines on Tesla—our grapes, and about twenty acres of vineyard recently planted on Mines. The property was most recently owned by an old Italian-family patriarch who died with no children; the vineyard was torn out by his nephews. There is no way to build or subdivide that remaining ground, and its highest and best use is in vines anyway. The nephews won't sell and won't give a long-term lease to someone who does want to plant it out. So, it sits, forlorn and unfulfilled in the middle of some of the greatest Cabernet Franc soil in California.

ii

When we bought this property back in 1997 it had been consigned, even while the family was living in it, to graceless destitution. The upstairs wooden floors were inexpertly covered in yellowing, cracked linoleum, and where the tiles didn't extend there were numerous holes through which a pencil could fall to the floor below. The back end of the house was tilted earthward like a conditioned beast that strove downward in advance of coming blows. Everything was worn and had a patina the color of the earth that would soon support a world-class vineyard. A crooked, heaving stairway led to the lower level. What a site! In about 2400 square feet, there were no fewer than eight small rooms divided up by particle board and bare drywall. I heard, later, that one of the owners had rented these small and dark spaces to passersby staying a week here or there.

There was not much worth saving. Over the course of the next 18 months, the house was stripped of all the forlorn garbage until only a few of its still strong roof beams were used to support a new upper floor. The rabbit warren of rooms on the ground floor was exploded out into one large expanse that became the cellar floor for a new working winery. The contractors lifted the mudstuck mass, dug down a few feet and built cinderblock walls upon which the rebuilt structure rests now. Instead of the wrap-around porch and Victorian inflected scroll work on the sides, one large stairway in the front was constructed as a grand entrance, the walls were stuccoed the color of an Italian sunset (one, in our imagination), and the dirt yard into which had been thrust a few miscellaneous trees, there is an expansive concrete patio. When the space was open to visitors a year-and-a-half later, there was nothing left of the former dilapidation except the ovoid circle of the drive leading one in and out. In the place of random trees and open dirt we planted 5.8 acres of grapes that we called our *Home Ranch* vineyard.

On the Home Ranch, which was planted 700 feet above sea level on the gravel leavings of the Arroyo Mocho river that ran through this area for millennia (the flow of the river carved up base rock into the finer chunks that eventually made our vineyard one of the finest Cabernet sites in California) in 1996–1997, to three different grapes; each ripens differently and is used for a distinctively different expression of authenticity and deliciousness. Sangiovese is used as a counterweight and counterpart in structure and flavor to Cabernet in our California version of a super-Tuscan offering called *Vincere* (first made for a friend named Marcello Fiorentino whose restaurant in West Palm Beach hosted, in 2000, one of the first winemaker dinners I ever did) as well as a 100% version of the variety whose greatest expression is in the wines of Brunello and Chianti.

One-and-a-half acres of Barbera butts up against the Sangiovese. I am guilty, like many of the early planters in California of planting varieties in my home dirt without thinking carefully enough if those grapes made sense in that particular spot. Sometimes the brand gets out in front of the grape and the brand message tortures the grape out of its most appropriate shape and into one that serves a predominantly money-making purpose. Barbera originates and thrives in the cool climates of northwestern Italy, in the hills of Piemonte and Alba. Northern California is significantly warmer than the original Barbera provenance. We produce wine that is delicious but will never have the truth about it that the original grape from its native dirt does. We search instead, as a by-product of our love for purity and authenticity, for a doppelganger of the wine that is true to its regional roots and true to the vision we have of the grape's potential for compelling, one-of-a-kind, wine.

Originality is not California's strong suit as far as wine goes. The same could be said, of course, of every region that is not Bordeaux or Burgundy, Chianti or Priorat (though one can historically argue that those grapes did not originate in Gaul or Iberia either). We have a couple of native grapes but none that makes fine wine. We imported those grapes (or more accurately, they were brought by the sons and daughters of the autochthonous lands where the grapes cleaved true to the soils and scapes and winds and culture of the originating place and brought them to their new homes.) The fact that California is about the most hospitable place in the world to grow things that you eat and drink did not hurt the chances of success for those vagabond European varieties.

We are still finding the exact right spots for each variety to grow, but part of that struggle is complicated by the fact that California offers such a large growing area that is not in the margins, that does not *hope* to be lucky enough to ripen fruit five years out of ten. California's challenge is abundance: of temperate weather, of ripeness, of richness. The seemingly innocuous conditions do a number, though, on the most historically accurate and most sensually compelling versions of the grape. Wines from my home state can be too much of a good thing. It is easy for the farmer to let sugar gallop ahead of the other ripening arcs like seeds, skins, and the textures associated with tannin. If you pick on sugar, in California's warmer growing areas like Napa, you get wine with high alcohol and jammy fruit but without the finer shadings of amplitude and modulation that mark complexity and elegance.

We use the seven tons of Barbera we get from our vineyard to make the wine of Alba and Asti. Wonderfully high in natural acid, Barbera is one of the quintessential food wines. Our picking decision, then, rests on acidity levels more than they do the levels of sugar. Balance in Barbera is a lovely challenge: to pick fruit when there is enough ripeness of sugar (and resulting alcohol) to bring the acid to heel. What we have found recently, too, is Barbera's affinity for producing fruit that is absolutely, unequivocally, inevitably, and (in all manner) perfectly suited to making a *méthode champenoise* version of the grape. Nancy Castro, a long-time wine industry veteran and my fiancée, and I drove down to downtown San Jose to meet my kids at a restaurant with a great Italian wine list. The list featured a sparkling wine of Barbera from Piemonte that we fell in love with. I thought we could make a version of the grape that rewards lovers of sparkling wine as well as one that hews true to the identity of the grape. I believe we have succeeded. Sparkling wine should bang loudly on the morning doors of the senses. It should introduce itself with an authority that is all debutante ball and hedonism. Sparkling wine creates the untrammeled day, and, by its very nature, leaves one open to the myriad possibilities of a set of experiences wonderfully shaded by the beneficent effect of alcohol. Wine with bubbles is celebratory. You could be one of those unfortunates that complains about winning the lottery; you cannot find a fault, no matter how hard you try, with well-made sparkling wine. It is a wine law.

Half of the Barbera fruit that grows on our Home Ranch is used for sparkling wine. The other half, about three tons, we make into our version of Italian Barbera. This wine, with its fragrant fruit and lovely generosity of structure, is one of the favorites of our wine club. There is an immediacy to this wine. The Bordeaux varieties need time. Hidden in the structure of the wine is an organoleptic truth that carries the hours like passengers on a slow ship, necessitating a long journey to blossom. Barbera is the opposite. It walks like John Travolta at the beginning of *Saturday Night Fever*. It has nothing to hide. To hide its nature would be to strip the wine of its fundamental reason for being. To hide its fundamental reason for being would be to rob the acolytes of that grape, in Italy and California, a medium of emotional and vinous expression that is essential, that is worthy of playing in the ballpark that hosts a world series of wine.

To talk about the Cabernet that is grown on the Home Ranch Vineyard is to encompass nearly the complete history of California viticulture. The first recorded planting of European cuttings in California occurred in the second half of the 1830s. Very nearly at the beginning, if not the first Cabernet vine shoved into the ground in California, the Livermore Valley received the largesse of Charles Wetmore, who was a true believer in the potential of Livermore and went to Europe with introductory letters from Louis Mel, a winemaker here, and a friend, too, of the Lur-Saluces family, owners of the renowned Sauternes estate, Chateau d'Yquem. Wetmore spent time in Europe in the early 1880s studying winemaking. He returned from his trip with a bounty of cuttings that represented the very best the Old World had to offer in Cabernet and the white Bordeaux varieties. Wetmore had previously visited Livermore and found, there, similarities of climate and soil with the Sauternes region of southern Bordeaux. Some of the cuttings Wetmore sourced came from the great estates of Château Margaux for Cabernet and d'Yquem for Sauvignon Blanc and Semillon. If one offers you the keys to the most valuable Ferrari to play around with as you see fit, you do not ask why. You hit the gas pedal and find freedom. Such is the case with the greatest plant material in the world. You plant it in your vineyards, take care of it, and make great wine from the resulting fruit. In fact, Wetmore did just this. After founding Cresta Blanca Winery in 1882, Wetmore entered his Livermore "Sauternes" in the World's Fair wine competition in Paris in 1889. Out of more than 17,000 wines, Wetmore's Livermore blend won top honors.

Concannon clone 7 is the plant material that came from those original cuttings brought back by Mel. Clone 7 is also the work-horse clone for

Cabernet in California. It is estimated that more than 70% of the Cabernet fruit that is harvested in California originally came from the Concannon cuttings, which mark their origin to the first-growth Margaux estate. The Livermore Valley stands in the middle of California viticulture. While rarely receiving the credit it deserves, Livermore is the incubator that has given California (and New Zealand's Sauvignon Blanc) world-class quality plant material upon which rests much of the total harvest of multiple countries. Clone 7 of Cabernet is planted on the Home Ranch and heat-treated versions of Clone 7 show up in another of the sites we use, Sachau Vineyard, as Concannon clone 8.

Planted in a rectangle with fruit positioned so that it gets the force of morning and afternoon sun, the Home Ranch contributes much of the best fruit from which we make Cabernet. While only 2.8 acres in size, the Home Ranch Cabernet is divided into three different subzones, separated according to the ability of the soil of each to hold water. We have learned a lot about this vineyard since our first vintage in 2000. Each of these smaller zones produces fruit subtly different from the other parts of the vineyard, and each contributes mightily to the complexity and world-class quality of the wine made from the site. Along the western boundary of the vineyard is the *Pepper Tree Horse Farm*. The name is not fanciful. Pepper trees (and Eucalyptus) were planted along the border of the horses and the vines to shepherd the winds that run rampant from the Bay. One of the subversive secrets of planting these specific trees is that they tend to dominate the terroir in which they are planted. Aromatic oils which carry the gorgeous stench of mint and pepper are carried by the prevailing winds onto the leaves and grape bunches that reside in that windborne orbit. Those grapes, from sub-block *A* are then harvested, laden with the oils of the neighboring trees, and all the stuff that makes its way into the red wine fermenter affects the final product in ways big and small. The long rows of zone *C* show great elegance and length while the short rows to the south of the site in sub-block *B* provide the heft and tannin for the wine.

Our cellar, then, holds barrel groups of the parts of the site that show the most herbal quality and the most classic California Cabernet nuances. Over the course of blending, we go through these separate small blocks, making a number of mock blends, in order to, eventually, reveal the wine that is the most quintessentially *Home Ranch*. Because of the world-class quality of the site, we are always tasting barrels with an eye toward their potential quality contribution to Lineage | Livermore Valley as well as to The Premier Cabernet Sauvignon, our finest Cabernet Sauvignon of the vintage.

The *Home Ranch* is one of our top three vineyards. What it lacks in dramatic, elevation-defying aspects, its nuances of soil provide an intensity of fruit and a balanced ripeness and roundness that please both the impatient (with its richness of fruit and integrated structure) as well as those who will allow the baby fat of the wine's fruit to age out, become more complex, and reward that cellar time with a truly profound wine that grows in intricacy every day and contributes to those experiences that grow richer with the passage of time. o

CHAPTER 8

THINGS YOU CAN NEVER REPAY

It's interesting how people come together. The circumstances of work, a chance meeting somewhere, will throw people together who should have no real affinity for each other. They don't see the larger world in the same way. But they do see slivers of the world the same way; they are both attuned to something very specific, and the connections are built because of that. I met my first assistant winemaker, Craig Ploof, in my tasting room years before he started working for me. He came in with his beautiful wife, Sagit, and seemed really interested in the wines. Looking at Craig, you wouldn't expect him to be a wine guy, and I think when I first met him, he was just becoming that. Goateed, ponytailed, seemingly fresh off an island in Hawaii, Craig didn't seem like the guy who would eventually help me make some of the best wines in California. We did hit it off, though, because of wine, and to this day, even though he has moved on, we will always be bound by wine, by the work it takes to make it, and the twin truths of passion and commitment that it takes to appreciate it when it is done well and right.

When I first met Craig, he was a glazier. He worked for a company installing mirrors and showers in upscale homes in the Tri-Valley area. He and his wife would attend winemaker dinners at the winery, happily putting up one night with my youngest kid, Sara, who got hopped-up on sugar packets and gleefully insulted Sagit's ex-husband, who was still friendly with her and Craig. They eventually became part of the family, joining us for Thanksgivings at the winery, and filling familial holes I did not know existed. Craig is an intense guy who sees a way things should be done and sticks to it. His work ethic is amazing, and I think, like most people who have to work their way up to the place they want to be, his capacity for work is held as an outward symbol of who he is. I grew to know this about him later, but for the first few years, he was a guy who'd come to taste at my place, who I'd see at other wineries and restaurants in town and who I got to know well.

In the late 2000s, when we had our Pinot Noir brand, La Rochelle, Craig volunteered to help with punch downs and racking and all the other scut work that is an essential part of making good wine. He'd come after his regular hours, in the afternoon, and he would help me, and we'd taste a lot of wine, talk a lot about the process of winemaking, and move through the days raising our wines up. On occasion, he'd be able to get out of his regular work, and we'd drive up together to a vineyard that we were looking to buy fruit from, and walk through the vines tasting fruit, and talking about whether the fruit was ripe or still needed more time.

In 2013, I hired Craig, officially, to be on the production team, and he ended up running the day-to-day operations of the cellar. As a vintage, 2014 dawned beautifully, and Aidan moved part-time from his hospitality functions to serve as Craig's cellar rat. The early fruit came in nearly perfect, and we had a couple of wines through fermentation and pressed off when my wife's first seizure came. Most of the rest of the winemaking year happened around me and my son. I remember being on a step ladder punching down boxes early in the morning with no one else around when I got a call from one of my kids that June was having a seizure. We lived in San Jose at the time, about 45 minutes south of the winery. I dropped what I was doing, all the peace of silence and sweat and early morning gone to adrenaline and traffic and visions of worst cases and then I headed home as fast as I could.

I met the ambulance at the hospital, and we found there was a stain of black on one of June's brain scans. The stain was cancer. It was taken out by a surgeon and found to be, in fact, the worst it could be. The rest of the harvest was a kaleidoscope of whirring and darkening shapes, displacing light and eating away the margins of normality until it descended, as did I, into an unsign-posted blackness. If not for my team, Craig in the cellar and Jennifer Fazio, Janice Fisher, Kathy Schoendienst, and Tracey Hoff at the tasting room, I would have pulled the brand down along with me.

There are things you can never repay. There is no currency for worry and loss and death and for making sure things function despite them. If you have a strong enough team, each member rallies, and they take on much more than the job description, and they captain a captain-less ship. Some of that team, including Craig, are gone, and the relationships have changed, as they must when it isn't just an employee and isn't just a job. Craig and Sagit opted out of sharing the holidays with us after Craig left, and just as I'm still unsure of Craig's reason for leaving in the first place, I don't get their absence from the November repast either. I'll never be able to repay him, and the others for that hard work, and for the feelings that he had for me, and for June, and for all of my children.

.... ¡¡ ....

When I joined Ivan Támas Winery on January 2, 1996, my dad and Ivan had already hired Jennifer Fazio, now the Vice President of Operations, who had worked for Wente Vineyards, our neighbors, (and partners for a number of years), first in their restaurant then in the marketing and sales departments. Because I was on the road two weeks out of every four and because we had opened a tasting room that needed to be managed, Jennifer assumed the general manager's role. Without her, especially during my late wife's illness, our business would be far less healthy than it is right now.

I think back on those early days of Ivan Támas and then Steven Kent, and I see a young and spirited team of people, like Mickey Rooney and Judy Garland, putting a *show* together. We mustered our energies to dig the post holes for the first sign on Tesla Road, moved a ton of wine around on a pallet-jack, painted walls, filled cracks, and did the myriad little things you do to get a business up when you have no money. We enlisted a lot of wonderful friends to help out, and the team pictures that we took at the end of the long "Harvest Festival" weekend each Labor Day, show a core group of folks who donated their time and friendship to help make sure the curtain actually did rise. Those pictures show, too, the growth of my family, the little kids sitting on their mother's lap, looking sideways as the shutter clicks; the awkward teenager, proud of the work done over the long weekend but also wanting to be anywhere but with a bunch of old people drinking wine; the photos of Aidan, now, still reflect the reticence to be center and front, but show, too, quiet pride in good works, and the dawning realization of his future role as the patriarch.

As I write these words, looking out the window of my tasting room at the waving vines, just before harvest, buffeted gently by the landward winds, I'm a bit melancholy. I've always subscribed to the notion that it is the day-to-day work that is most important, those long days of toil from which you gain your inspiration for the next day, and not the destination that you've either talked yourself into or envisioned after long years of planning. The daily work is guaranteed; there is no endpoint without the work that is circumscribed by the rising and the setting of the sun (and this is what happens when you are deep in harvest… you wake up when the orb has yet to breach the eastern mount, and leave when it has descended, molten, into the western waves). What is not a surety is any success one may have hoped for, much of which is beyond one's control. So, you do the work. One day after another, punching down, walking mile after mile between the green-curtained rows of vines grabbing samples, thinking and tasting and blending, working hard to make something beautiful from unruly nature and offering a gift, at the end, that you are not sure anyone really wants. It takes a special dementedness to find succor in this gentle chaos.

I wish that we were further ahead, that we had made more cracks in that ceiling that separates us from the vinous denizens of more famous growing areas, to be sure, and I am competitive by nature. As much as I'd love a sense of financial stability, I have grown to realize that much of what passes for success is beyond my control. This realization irks me every day, but it does not lessen my pride in the team we are building, how far we have come, the quality of the wines and experiences we are creating, and the fullness on the horizon ahead. ○

PART Two

"It is the place where passionate inquisitiveness meets the grumbling stomach; the place where you ask that girl you love to eschew all others, the place where business successes are celebrated, where simple laziness is excused and where pods of people come together as they did around some ancient fire, in communion and brotherhood, to celebrate the most mundane and important of things: a full stomach."

CHAPTER

9

THE TRUE NORTH OF CIVILIZATION

In July of my 11th year, we ventured out of San Jose for Lake Powell and our annual trip to that amazing place. My dad was the VP of Sales for the family winery, so he was gone most of the month, traveling the country, introducing one wine market after another to the quality of the family wine. My dad loves wine. He has a passion for the process of making it but even more so for wine's ability to convey and encompass a whole world of emotion and friendship at the same time that it is a convenient post upon which to pin memories of experiences and people now past. He met a lot of people along the way, and his genius was the ability to make friends with everyone.

    There was one family, the Doerschlags, that I couldn't wait to see each year. I was too young to fully understand the glories of its daughters, but I remember being entranced by the older son. He was probably five years my senior, and I recall 45 years later how smart he was and how much attention he received from the adults. He ended up a geologist, I think. He did know his rocks. Mark had this preternatural self-possession and an ability to walk among the adults in the room without pretense and without doubt. He'd order basted eggs in the diner and know the origin of every geologic formation the rest of us would see simply as the scaffold of our summer fun. This weird kid who knew all the answers played a prominent role in my early life. He, maybe more than anyone (including my father), taught me what it was to truly eat.

Our drive to the lake would include a stopover in Hanford, California, most years. I'm sitting at my desk 45 years later, listening to Miles Davis in the background, writing *Hanford, California,* as if these words were merely descriptive of a Central Valley locale, atrociously hot in the summer, devoid of any sense of personality, bearer of tumbleweed and flaked paint, sweat-raiser, sun-strained and sun-struck. The small and inadvertent surprises of life are the ones that seem to have the greatest effect on who one becomes. Big stops produce obvious results. The less obvious qualities, the ones that abide in that incorporeal part of each person, and the ones that explain all the most vexing of action and thought, sometimes lie dormant for ages, ticking away at their own pace, laying just at the margin of the mirror. It is in these shadows that you glimpse only obliquely, where you realize (if you're fortunate and open to such possibilities) that the deepest and most authentic part of you stands ready to be mined.

Hanford, California, was home to the *Imperial Dynasty* restaurant. Owned by a gentleman named Richard Wing and consisting of two sides and two cuisines and two philosophies, the restaurant served as the sperm to every esthetic contemplation I've since had. The scene in the *Wizard of Oz* when Dorothy emerges from her careworn Kansas house after landing auspiciously and fatally in Oz most perfectly describes that cinematic trope of irreparable division. Between youth and age, innocence and cynicism, hope and the zero, there is that agent of change, most times, especially when one is 11 and tired and hungry from a 15-hour ride in the back of a tricked-out van, the moment that one mutates from blind to sighted is never fully realized or only much later, if you are really lucky.

.... ii ....

I ate out a lot as a kid, especially when my sister and I were spending our weekend with my dad. *Renzo's, The Round House, the Crow's Nest, Paolo's* all bring out wonderful memories of the lively din of clattering dishes, the silent ballet of waiter and busboy and maître d', the restaurant owner greeting my father warmly, and (mostly) the textures and flavors of the food; the beauty of animal reimagined as entrée, the sense of camaraderie everyone in the restaurant tacitly shared, and the lifegiving-ness of the vibe. When I stumbled into the *Imperial Dynasty* for the first time in 1975, seeing the carts and red calendars and fringe-festooned waitstaff, I was comfortable in my novice's understanding of Chinese cuisine. I think the food was good though I never ate there.

In an adjoining room, through a curtain or a door, I cannot remember which at this point, a secret garden opened before me. Instead of the lushness and rarity of night-blooming flowers or the ordered hedges separating one from the unruly world outside, there were dazzlingly set tables (with so many forks and knives!), waiters carrying wine around in large crystal decanters, the shells of snails bathed in green-flecked butter, the exotic aromas of fish and fowl and meats that I could not identify, luxurious as meringue, set upon the tables, and the restrained joy of the owner, Mr. Wing, when greeting my father. I had entered an entirely new place, one that begged decoding and surrender. Fantastic!

Richard Wing had come from China when he was a boy, settling in the middle of California because there was work there in the fields for recent immigrants. He had learned how to cook from his grandmother, the traditional dishes of his home, but went on past the familiar to the cuisines of Europe because he wanted to make people happy. He enlisted in the Army after Pearl Harbor and became General Omar Bradley's private chef in the theaters that had previously trained him as a young man but that now bore the scars of mortar and desperation. After the war, Wing returned to Hanford and opened up a restaurant catering to the un-ordinary appetites of those locals who came to love the mock exotic *chow mein* and *moo goo gai pan*. Unbeknownst to most of the local patrons, there was a true world of exoticism just on the other side of the thin velvet-covered door. Though I knew my family was in the wine business and that my father went about trying to sell it, it wasn't until I went with my dad and Richard down a steep flight of stairs to his basement wine room that I began to understand the relationship between food and wine, and the joy one experienced descending those stairs to a secret room with the owner of the restaurant as your guide to find, like Howard Carter at King Tut's tomb, "wonderful things." Many years later, as an adult, devoting his life to the business of taking care of people through wine and hospitality, I came to regard this relationship of wine to food in great restaurants to be connected in wonderfully complicated ways. Wine and food exist on strands of the same large web. When the chef pulls on the Normandy strand there is a sympathetic movement on the strand that is cow in meadow, amazing dairy; the Ligurian strand dances with the seafood of the Amalfi coast and the strand that supports light white wines that express the briny sea through a prism of fresh fruit and acidity. There are family history strands, the food-of-the-chef's-childhood strands, the natural-bounty-near-the-restaurant strands, and the typical-grape-of-the-region strands. While all the life experiences that go into making a chef impact her food in ways subtle and obvious, it is the relationship of her food to fine wine that transforms ingredient into meal.

The food of the great chefs is a painstakingly thoughtful and radically imaginative distillation of centuries of technique and training upon which the empires of millennia-past forged the most enduring portion of their legacy. While these chefs, or at least the product of their intellect and emotion, spread as some integumental diaspora, creating fortresses of culture for worlds new and original, it was the bottle of wine, hidden in the darkness of the cellar below, that most distinctly connected the food eaten to the one who ate the food. The indescribable genius of great Burgundy reaches its apotheosis when paired with the most delicately transformed duck; the duck, too, reaches levels of *becoming* way beyond the significant inputs of and entanglements with great wine only. The web and its individual strands, grand cuisine and ethereal wine, *communicate* with each other, make synergies beyond the ken of human intellect, to reveal an ecology of taste and experience that connect one to fundamental things.

In Wing's cellar were the great wines of the world...the finest Bordeaux—First through Fifth growth—and the Premier Grand Crus of St. Emilion, the profoundest domaines of Burgundy, Chianti Classico, California new and gone. There were a multitude of shelves full and stacks precariously balanced—of bottles, green, flint, and brown. In a diamond-shaped wooden enclosure that housed 12 bottles of wine was a *Mirassou* Cabernet from 1968. The wonderfully old-timey white label with the foothills stolid and green on the lower border, the *Mirassou*, in a fancy black script at the top, was instantly familiar to me. This wine was a blend from the original home vineyards of San Jose and fruit from the recent plantings in windswept and unwelcomingly cold (to this Bordeaux variety) Soledad in the heart of the Salinas Valley. Six months ago, and many unsuccessful older bottlings from my father's generation later, I would have scoffed at the idea that these wines were ever any good. Too many stories of the incompetence of one family member or the other and my own hubris precluded me from believing in the potential of what was actually being crafted. I had the opportunity to taste the 51-year-old version of that '68 Cabernet, and I marveled at the vitality and flavor that still remained. The second bottle was even better. Sometimes, we allow personal feelings to get in the way of a more inclusive and connected wine experience, much to the detriment of the winemaker and the wine and the drinker. I appreciate having my prejudgments contradicted so thoroughly. What is life without growth?

My father and Richard talked about great wines that they had tasted, previous experiences they had together at the Imperial Dynasty, the state of the hospitality business, what was on the menu for that night's dinner. We came up out of the cool cellar, my father's arms full of a selection of amazing bottles, into the light of the magical dining room, and it was like being birthed into the Garden. Over the next several hours, I experienced one of the finest collections of textures, aromas, and flavors—organoleptic virtuosity on a heretofore unimagined level—that I have yet had. Like Dorothy Gale seeing a universe of color upon reaching OZ, my own inner, technicolor portal opened; I would never be able to see food and service and hospitality as mere work or stomach-filler again.

The restaurant (from the French verb *to restore oneself*) as we know it now is only a few hundred years old in the West. It wasn't always that the weary traveler, coming in from a long night of riding, wet from the falling rain and chilled from the winds blowing across the darkened landscapes, could find a place of solace, a place to be cared for. There came to be a profitable need for these waystations, these gustatory hospitals, and there came to be a class of purveyor who carried within him a connection to the repasts, both current and of the grandmother, that claimed a cultural importance and permanence. These meals, though prepared by strangers, united the diner with his own past while also introducing him to the most forward-thinking ingredient and preparation. The restaurant grew from the place where sustenance could be gotten to the place where food and food preparer were celebrated and fetishized. Ranking systems of quality grew out of a period of relative abundance, and with the advent of a million stations on the TV, the chef became the subject of hagiographical biography while the restaurant was reduced to a prize for the seeking and unfulfilled diner. The purity of the original vision seems to have been lost in many of these temples. The fundamental mission to care for a stranger, to make him feel a little less foreign, to add joy and richness to his life, often seems lost in the striving for transitory recognition. The restaurant is the true north of civilization. It is the place that spawned the American experiment, harboring Hamilton and his cohorts as they made revolution. It is the place where passionate inquisitiveness meets the grumbling stomach; the place where you ask that girl you love to eschew all others, the place where business successes are celebrated, where simple laziness is excused and where pods of people come together as they did around some ancient fire, in communion and brotherhood, to celebrate the most mundane and important of things: a full stomach. It is no matter the kind of food served in a place, the décor, the wine list, the popularity, the Michelin stars—all of these are filigree. It is the commitment to care by the owner and the staff that makes a restaurant true.

The restaurant is the place where the hurly-burly of the day is quarantined. You enter alone, infected by the tumult of the noisome and uncaring world outside, you are made comfortable and are fed (an act of intimacy second only to sex), and you find yourself cordially bound to a tradition of work, horrendously difficult and so glorious because of that difficulty, that connects you to things beautiful and delicious, sustaining and transitory. For most restaurateurs the transaction is not a financially equitable one, and the attrition of restaurants is staggeringly high. If not for the money, why do these lunatic and magnanimous people do what they do? Simply, they must. And we are, very fortunately, the recipients of their compulsive caring. ○

CHAPTER 10

THERE IS SOMETHING SACRED

Three of us are in the lab at the end of a long day of tasting, and we're writing our notes on the flight of Cabernets we're tasting through, going from glass to glass, tasting, spitting, evaluating. Aidan comes from around the other side of the table to take the eighth glass in the flight, I'm marking my grades on the table with a marker, and Beth is typing notes on her computer. In the background, music is playing from Spotify on a speaker, and the lab folks, who work for Wente but who also run our numbers, are packing up their stuff and going. I return to a glass that I'm not quite sure about and grab a few pistachios from a bag that Nancy has left for us because I haven't eaten anything since breakfast, and no matter how much you spit, you still get a little sideways.

I'm not sure how I came upon this very specific organizing vision when I taste, but I cannot get this picture out my head. My mouth is a cave. My cheeks are the walls, my tongue the stream that runs through it. This cave and this stream are transformed (or one might say, created anew, with each swallow of wine) by the physical nature of the wine that enters it. I take a sip and immediately I am thinking about how the wine's acidity speeds up the flow of the river or causes it to meander, drought-stricken, finally into a quickly drying puddle. If the river is running, is it full of the white-foamed undulations of tannin or moving, serpent-like, with all its motion under the surface? Is this cave, which is housing this mutable flow, cavernous or simply a small rent in the side of an adamantine mount? If it is wide and open, then the tannins and mouthfeel of the wine are felt all over my palate… the cheeks, roof of my mouth, back of my throat. The opposite describes a wine that is very soft on entry, one with too little acid and too little structure. The length of the cave is related to the urgency of the river's flow. The greater the acidity, the longer one senses the wine in one's mouth, the longer the cave through which the river runs. This perception of tasting seems overly complicated to me and smacks a touch of one suffering from synesthesia, but I cannot banish this strange picture no matter how casual the current experience with a given wine may be. It has served me well, strange as it is.

The day is winding down and three people are doing what they love, and I even say as much… "isn't this fucking amazing what we get to do?" We're about to talk about the ten wines in front of us… what's compelling about #2 or why we should kick out #7, and we are each—the three of us—focused on our mission and in love with what we do, and a little weary, but only from standing up all day. We each came to our craft with different experiences and different goals, but now together only one thing matters. We discuss the structures of the wines, defend the wines we think should be included, make the case for eliminating the wines that aren't good enough. We three are there in that room, the echoing of equipment outside in the winery winding down at the end of the day, tasting those wines, together; we three, fighting off the distractions and fighting off the fatigue and the accretion of tannin on the tongue from the hours of tasting. We three, with the future of the brand at stake, late into the day, together. Hungry and tired and thirsty and in love and devoted and… brothers.

Throughout this process I am keeping in my head a picture of the wine I ultimately want to make. And each barrel I am tasting is evaluated in the context of whether it is a piece that can fit into this disarrayed jigsaw puzzle, filling in a hole or sanding off an edge or whether it is a disconsolate piece that fits in perfectly to another wine, making that next wine something more fine and more whole. By the time I have finished, my finest wine will have been made by including then discarding, discarding then including, the best 16 barrels out of 200. My team and I will have agonized, finally, over the last Petit Verdot barrel or the last few gallons of Malbec until we have in the sample bottle at that moment, the best wine we can make for that season.

I have gotten a great deal of joy over the last couple of vintages working with my young winemaking team. Aidan and Beth, because of their youth and inexperience, bring a sense of discovery and joy and wide-eyed openness to everything we do. Every time we are in the lab blending or tasting a wine that they have never had, a palpable thrumming of excitement emanates from them. Having young people around you is the best way to stave off the *been-there-done-thats* that intrude when you think you've seen everything. The time we get to spend tasting and blending together, being there as a team, at those moments when we are a-wash in that peculiar energy that is a by-product of *making* something is as religious an experience as I can have.

One of the consequences of creating something so evanescent, yet so emotionally enduring, is that the whole process is shaded heavily by the temporary and mutable reality of the wines we are putting together. Each time we taste, we taste differently. Maybe we didn't sleep well, maybe we didn't have enough water or ate something weird (for me, it's apples) that affects our ability to taste. Attitude is crucial for success. If I am worried about home life or cash flow I am not fully engaged in the process. It is this full-throated connection to the wines we are working with that suggests at least the potential for success. The tenuousness of the blend is beautiful and daunting. We only have one chance each year to make Lineage or The Premier Cabernet. If we fuck it up, there is no recourse until the next vintage. There is a tremendous amount riding on each wine and on the process by which we arrive at that vinous conclusion. We work in a realm of fluctuation, of momentary perfection, a realm that is glorious because of its fragile and evanescent core so we have to be comfortable with the lack of surety, of the transitory nature of each wine, no matter how long-lived it might end up being. If we can continue to weave together a long-enough string of successful wine releases (those that speak of authenticity and deliciousness and a perennial blossoming of the relationship between nature and man) and be comfortable enough walking along the fine edge of success, we will have created something truly magical.

There is something sacred to the act of tasting wine like we get to do. Tasting wine is different than drinking. Not for the obvious: we spit when we taste, and we are thinking deeply about how that particular wine helps to finish the picture in our heads that is the combination of disparate parts and the culmination of years of effort. When we are drinking, we are coming out of ourselves and sharing that social part of ourselves with the others in our group; the wine is sometimes superfluous to the connections we long to make, even if we don't know it at the time. When we are tasting in the service of making, we are connecting the most intimate bits of our profession into a cohesive picture that becomes a cohesive story at the end. If we have succeeded.

And it is not that our story spans only a year. Each act we perform, the harvest, the pressing, the blending, the bottling, is connected to each other like pearls on a string, but tied together with all the other times we, ourselves, have done these things, linked, also, to every other time these things have ever been done. The pearls are the single tasks of this year, joined. Those same pearls are the single years that go to make a career just as they are the countless winemakers and the countless wines and wineries, unnumbered and unnamed, that have worked to draw out beauty from oblivious Nature. To truly taste a wine is to taste the dust of the past, the failed blends and faulty visions; it is to sip from bitter vials and know the ache of the black and ruined season, the fires that ravish the hillsides, feeding off the ancient trees, bringing smoke and ruination to the vintage; to feel the wind driven forth by the advancing storm and to see in the blackening light of the onrushing clouds, the rain and the hard hail that chops through leaves and fruit like a hatchet. To taste truly, as one node on a giant esthetic network, is not only to gather deeply what has been offered but also to commune with those brothers who despair at the loss of a vintage as we celebrate those who've made their best wines. To drink of the effort and the management of fear and the quest for oneness and the quest for excellence and the blind hope and the fatalistic acceptance with comity and empathy is to stand on that one threshold together with all those other winemakers, together and brothers, in one never-ending chain. ○

∞

L to R: Aidan Mirassou, Beth Refsnider, Steven Kent Mirassou

∞

CHAPTER 11

GODS OF ASPHALT AND SHINGLE

I started working at the family winery when I was 12 or 13. I'd ride in each summer morning with a guy who worked for my father in the marketing department and then ride back with him too. The winery buildings on the east side of San Jose, an area called Evergreen, had been built very early in the 1900s. When you'd drive in past the administration building next to which you'd park, you'd pass by the scale house and over a huge metal platform that would weigh-in all the fruit coming in from the vineyards. There'd be a very distinctive bass *chunk* as your tires left the pavement and alighted on the floating wedge and a reverb of sorts as you got back on to the pavement on the other side. I remember very distinctly, that first summer, how mysterious everything seemed, and how freighted with a sense of importance and magic each small action had.

I remember each morning started with the timeclock ritual. You'd run your finger down the cardstock timecards looking for your name, you'd notice the blue ink, often smudged by the clock mechanism itself, (so many thousands of *kerchunks* over the years), the small 8 and the even smaller 00 superscripted above it. Each day, in and out twice—once for lunch—you'd put the card in, and you'd get your days marked off. My card never seemed to be in the same place I left it the previous night, and I'd have to riffle through all the Spanish and Filipino surnames until I finally got to mine. The mysterious innards of the family business were entered and engaged through this tiny waystation. I am sure for the folks that had been there for many years this place carried no more meaning than the average closet in one's home. To me, though, at that age, it was trapdoor and secret-agent stuff.

One summer I worked in the Champagne cellar on the disgorging line. We had a couple of large warehouses on the property; one, the NC warehouse was where we stored all the wine that had been bottled and was waiting to be shipped out. This huge space was named for my grandfather, Norbert Charles Mirassou. The EA warehouse was where our sparkling wine was made, aged, riddled, then disgorged. Edmund Aloysius Mirassou was the patriarch of the family, my great uncle, and the namesake for my summer workspace. Ed was my grandfather, Norb's, younger brother, and was the marketing and administrative whiz behind the family's brief ascendancy in the 1950s. I didn't know him well, but I do remember him to be a force, serious, yet kind to the kids of my generation, at least those on my side of the family. Ed had three sons and a daughter; my grandfather had one boy and one girl. Back then only the boys were invited into the business; the girls were well-compensated later but were seen as superfluous.

I remember the EA warehouse to be a dark place at the margins where some finished wine was stored in cases on the periphery of the building. The small production space, carved out of the larger expanse, was well-lit, I remember, almost as if it were a stage. The pallets of wine off to the side resided in the darkness like a rampart, and their bulk, unseen, was suggested, really, by the way sound died at the edges up against them. The bottling line was set up in a U-shape, and I stood mid-way on one of the arms adding wine to a predetermined mark after the bottle had been disgorged.

Champagne (called Sparkling wine now in most parts of the world except Champagne) is a magical thing. The legend is that a monk named Dom Perignon discovered the process for capturing carbon dioxide in bottle, thus creating a style of wine that is known the world over as the beverage of celebration and victory. It is certainly this. It is also so much more in that it may be the wine most dominated by its texture and its relationship to a gaseous world that is never contemplated in still wines (unless there is a flaw) but that contributes mightily to the age-worthiness of the wine and how it lives and grows before its eventual and inevitable decline. Champagne starts life in the same fashion as most other non-dessert wines. Traditionally, a blend of Chardonnay, Pinot Noir, and Pinot Meunier (the *Miller's Pinot*, so named for the fine white hairs growing on the underside of the leaf that resemble, from certain angles, the flour the miller mills), these grapes are pressed and made into a wine with very high acid and low alcohol. So far, so good. This wine could be drunk as is, though it would be very racy and would lack the richness that fully ripe versions of these grapes would render. But this is part of the point. This base wine is only the first waypoint on the fermentation express. As we know, carbon dioxide is one of the by-products of fermentation, and the genius of the first practitioners of the *méthode champenoise* was to realize that if a second fermentation was coaxed in the bottle and that CO_2 was captured in the wine, one would have an epiphanic creation hitherto unknown in wine. The Dom Perignon was reported to have said, when tasting sparkling wine for the first time, that he was "drinking stars." Regardless of the veracity of the story of the Dom or whether the method of Champagne produced the first sparkling wine (many believe the *méthode ancestrale* or *Pétillant Naturel*, one in which the wine was bottled in the cold of the autumnal season, before the primary fermentation was complete, and the resulting bubbles were the result of the captured carbon dioxide as the still sugary wine woke up with Spring warmth and finished fermentation), for the purpose of the limning of the fourteenth summer of our protagonist, we'll stick to it.

So, to the first wine—the base—the winemaker will add a small amount of yeast and sugar to each bottle before it is adorned with a crown cap (the wonderfully ordinary bottle cap). Again, as we know, yeast love sugar, consume it voraciously, leaving behind alcohol and CO_2. Because of the crown cap, the out-gas is captured in bottle and forced back into solution. The resulting pressure of all this sequestered gas is seven times greater than that of the atmosphere we breathe; it is also the thing that generates that viscerally satisfying *pop*! when the cork is finally pried loose of the bottle. (It is also scientifically known that this sound is the third most pleasing to 97% of human ears).

The single-celled yeast not only perform their primary role of sugar transformation, they contribute that quintessential organoleptic matrix of yeasty biscuit as the increased alcohol concentration in the wine literally explodes the little beasties. As much as any other quality that this style of wine displays, it is the textures, aromas, and flavors of yeast shrapnel infusing the wine over the years as it rests *en tirage* that are most recognizable. There comes a point, finally, where the winemaker determines the wine is ready to be finished. In still wine production, there are a number of occasions in which the wine is decanted off the yeast cells—the *lees*—and clarified before bottling. Disgorging is the explosive method used in sparkling wine production to create clear wine. A few weeks shy of *dégorgement*, bottles of wine were historically put into A-frame racks punched through with the holes that would support those bottles. The wine would be put into the holes perpendicular to the A-frame. In those few weeks, a cellar-hand would turn the bottles slowly, lifting each one gently, until it stood nearly vertical in the rack. The dead yeast cells would ride the gravitational wave downward to the neck of the bottle, leaving behind clear wine above its grainy plug.

That summer when I was 14, I would go every morning with a bucket in hand to a valve located underneath the flying catwalks of metal that cut the morning sun into triangles and parallelograms and that linked one 30,000-gallon tank to another, and I'd get a full pail of glycol—antifreeze, basically—and I'd bring and dump it into a metal tray full of dry ice. The tray was about 6 inches deep and on wheels, three feet off the ground. On top of the tray was an A-frame cut in two, so it lay flat. A Filipino guy named Jesse would fill the A-frame up with 24 bottles and the necks would be submerged a couple of inches in the glycol and dry-ice. It only took a few seconds to get the neck of the bottle and all the yeast and schmutz that had migrated down during the riddling process frozen solid. At Jesse's station he had the bottles on his left side and a metal box standing in front of him with a little tray to lay the frozen bottles on. He had a modified bottle opener that the machine shop had made, and he'd yank the metal caps off inside the box. The frozen plug would fly out, propelled by seven atmospheres of pressure, more than one hundred pounds per square inch, which would bang noisily into a metal bucket laid on its side in front of him. The wine would bubble over a bit, and Jesse would place the disgorged bottle on a metal conveyor belt the width of one bottle, running slowly. I was the next one on the line. I stood perpendicular to him staring at a diffuse light box across which was taped some kite string at a height that represented a full bottle...750ml. It was my job to make sure that the bottles were full up to string level. I had a rubber hose, the same kind we'd make sling shots from when we were younger kids, with a clamp attached to a small tank that contained still wine in which was dissolved a very small amount of sugar. The *dosage* was calibrated to create a finishing level of sweetness in the wine. It is the dosage (later, when I had a Pinot Noir brand called La Rochelle and we were determining how dry we wanted our sparkling wine, we would do a *dosage* trial, adding ever-increasingly increments of the sweet *liqueur* to sample bottles of bubbly wine until we were satisfied with the balance of acid and sweetness) that makes a wine a *Brut*, or *Demi-sec*, *Extra-Dry*, and so on. Then, at 14, all I knew was that I had to look through the Champagne-green bottles, illumined by the egg-bright light behind them, add the dose, before they went off to be corked and caged.

 Eight hours a day the line would move, clattering continuously, the yeast plugs were ejected with the force of gun shots, and the finished bottles would bang glassily into cardboard boxes. I remember Jesse, tall with a goatee and a slight Asian cast, Vivian, the boss, also from the Philippines, and Kim, a Korean, always with a smile on her face. We were the group, for that summer, that made Mirassou Champagne (it was still legal to call it that then) ready for our customers to drink. It was easy for me to fall into the rhythms of the crews back then. I am a natural people-pleaser (much to my chagrin, if I think too hard on it) and was more so at 14. I am also prideful and pathological about pulling my weight. This was a big deal for me as a kid, especially at the winery. I had a bunch of cousins from the other side of the family that weren't worth spit that would occasionally be hired on during the summers too. When you had the name, it wasn't like you had to have a sparkling resumé. I'd hear the real workers talking about my cousins, about their relative worth as laborers, and I made a promise to myself that they'd never have cause to run me down too.

I loved being part of Vivian's crew. She was tough and would give me shit if I deserved it. We'd take a break around 9:30 in the morning and walk to the break area on the other side of the plant, and I'd grab a grape soda out of the machine, that old kind that still had glass bottles, ice-cold and delicious in the heat of the day. The crew would stick together as we walked under the catwalks that my dad had built when he was twenty, filled now with the pump crew, some who'd catcall good-naturedly to the women below. It didn't take me long to lose my youthful and spastic enthusiasm. I assumed the same laconic, ironic, and cynical walk as the others of my team came by naturally. If I had been a smoker then as a teenager, I'm sure I'd have a cig hanging out the corner of my mouth, smoke puckering one eye shut, as I moved wearily from bottling line to lunchroom. This was an act though. I'd play it to the hilt if we had to walk through a tour group on the way, as often happened. But in reality, I loved being a small part of what this band of brothers accomplished, and my feigned show of cool was a salute to their steadfastness and hard work.

Other summers I worked with the shop guys going out on calls to fix equipment and vehicles. My boss there was later found to have embezzled hundreds of thousands of dollars from the company and eventually went to prison. He treated me okay though. I worked a summer with my first cousin, Mike Alexander. His mom and my dad are brother and sister. Mike is one of the great people I know. He was 18, I think, this one summer, and I was 13. There were a couple of gas pumps planted near the conference room that our fathers used to meet in. The family and the managers of the company could fill their cars up and the company cars and trucks would fill up there too. This was back in the 1970s and the safeguards in place at even the most disreputable gas station now were light years ahead of where we were then. My family had a lot of smokers, and it would not be unusual to see one of them, cigarette in hand, pumping his car full of gas.

One of our jobs was to build a metal railing around the gas pumps so someone wouldn't accidentally run a car into them. I remember Mike and I measuring how long a piece of iron we'd need and then getting the shop guys to build an "L" that we would cement in place. We had a jackhammer to get through the layer of asphalt and some picks and shovels. Mike ran the hammer, and I shoveled the shit away from the growing holes. If we knew how many times we came this close to calamity in our lives, we'd never get out of bed. Lucky for us, Mike and I were not graced with this Cassandrian vision. Not knowing what we were doing or where exactly the underground pipe that carried gas from the holding tank to the pump was located, we merrily went about jackhammering the holes into which we were going to place the "L" bar. We realized just prior to disaster that the tinging sound we were hearing was Mike banging the hammer into that fuel-filled pipe. I can remember him shaking his head saying, as we were mixing up concrete to fix the railing in place, *are you fucking kidding me*, over and over.

Mike was also the one who taught me *Winese*, the language of hormonally-addled boys riding the metal highway above the tank farm. Often, during the summer, there would be tour groups roaming around the winery grounds after tasting wine in the Tasting Room. I'd be going about my various duties, washing out fermenters and storage tanks, weighing grapes, working on the bottling lines, and I'd turn and nearly bump into a crowd of people stepping gingerly over uncoiled two-inch hoses, pushing in for a closer look. All the work, no matter how mundane, is fascinating to someone who has never seen it before, so I get the attraction. For the guys, like my cousin, who were in their late teens and early 20s at this time, the tour groups were a beautiful distraction. They'd call to each other, like birds in the forest, singing the praises of some particularly attractive girl: "Lud-O-O-kud//A-tud//tud-hut-A-tud//hud-O-tud//O-nud-E//O-vud-E-rud//tud-hud-E-rud-E!" Winese is a spoken language so it doesn't scan particularly well. Consonants add -ud to the end, and vowels are pronounced like the long-form. So, Hud is "H" and "o" is pronounced "oh." Then Tud is a "t." I first learned the language on a drunken night on top of a houseboat in Utah decades after the fact. I pointed to my stepmother and said "so Lynn is Ludd-Yudd-Nudd-Nudd?" She did not find much humor in this translation the first time she heard it. To this day, the Winese translation of "Lynn" always makes me smile.

The summer I worked in the EA warehouse was the only time I got in trouble at the winery. The family has always made really good sparkling wine. The original winemaker was a German named Max Heubner who came to work for the family shortly after the second World War. He was a terrific guy, always treated my sister and me well. More importantly, he was a great winemaker. Mackie, as he was known to us, would make a variety of different styles of Sparkling wine from Chardonnay and Pinot Noir and would experiment by adding different flower petals into the fermenter as they were going through the primary ferment. His wines helped to polish the family name. One day, at my lunch break, I looked through the conference room windows and saw a line-up of some of the best Champagnes of the time intermixed with bottles of our wines.

My father and his cousins had been tasting these wines and evaluating our efforts against the best competition in the world. I was with my cousin, and we went into the room and proceeded to taste through the string of bottles. I'm not sure how reliable my memory is regarding my intent, but I don't think I went down the line hoping to get a little shitty. I think I really wanted to know what these mysterious wines tasted like. I didn't spit (as I would now), and I think that may have given me away. I was called into my boss's office the next day, and he proceeded to tell me how irresponsible I was and how I could have jeopardized the family business if I had been seen by the wrong people. Well, the *right* people ended up being bad enough, as the guy who ratted me out was one of the assholes that had been stealing money from the family. Despite getting reamed-out, I remember those wines fondly. I will never look upon a Taittinger *Comtes de Champagne* or Roederer *Cristal* with indifference.

The winery was a place of magic for a young kid. My younger sister, Pam, and I would run around on the equipment on the weekends and grab tabs of unmelted glue from the hotbox on the bottling line and chew it like gum. We'd make forts from cases of wine stacked on pallets in the warehouse, and clamber along the catwalks outside connecting one huge storage tank to another and that gave out on amazing landscapes both natural and man-made. From these perches, you could see all the way out to the base of the eastern foothills, empty except for the acres and acres of apricot and cherry trees that gave Santa Clara County the moniker: *Valley of Heart's Delight*. The lab was full of equipment whose uses could not be fathomed; the redwood uprights were heady with fermenting juice.

Our tasting room usually had a fire going in the great stone fireplace (made by a contemporary friend of my father), no matter the season.

Above it, 12-feet off the ground, hung the head of a boar my maternal grandfather had shot in Hawaii. On the walls hung the stolid, disappointed visages of the five generations that preceded me (those same pictures can now be found hanging on the walls of my Tasting Room in Livermore. Generations 1–3 do not seem any more content to be hanging more permanently in my room, however). No matter my age and no matter the time of year, cold, echoing rooms, the dark storage containers, funky-smelling drains, and broken glass always seemed to have a warm and glowing light illuminating them.

Thirty years later, when my cousins decided to sell the land upon which sat the Winery, I climbed up those metal steps again and looked out on those same acres planted to housing development and strip malls. Some of the best moments of my childhood occurred in and around those orchards (I remember holding tightly to my father as he plunged us full out through those trees on his Kawasaki. I learned to spell *Autumn* in the dirt of our picnic spot, forgetting the *n* the first time. Coincidentally, I won a second-grade spelling bee with that word a few weeks later, and nearly 50 years after that my oldest daughter gave my grand-daughter that same sonorous name). It wasn't long after that the orchards were removed one section at a time as new streets and houses were cut in to accommodate the burgeoning, high-tech work force. The long and ineluctable march toward progress, at least as the developers saw it, left a scar on the land. The carpet of white blossoms that bloomed in Spring and covered miles of area around the winery, throwing off a marvelous scent of rebirth, was sacrificed finally to the gods of asphalt and shingle.

My last summer working at the family winery was my sixteenth; the following year I would spend that time on a Eurail pass I'd paid for from the money I made slinging pizzas the winter and spring before. Until I sat down now and thought about those summers again, all these years later, and had the thousands of smells and sights and words and deeds come back to me with complete clarity, those days seemed lost in the haze of all the days, consequential and not, that have wound out since then. Each of those discreet packets of time had, I realize now, a small momentousness to them. We are products of all the experiences we accumulate to be sure, but not all of them have equal weight in determining what it is we will do with our lives. Those teen-age days, that I so willfully and overtly denounced when I went off to college, won the day, it seems now. How fortunate I am!

When the Mirassou brand was sold by my cousins in 2002, they still maintained ownership of the winery facility in Evergreen. With the family brand gone, they created a new mark called La Rochelle that specialized in wines made from Monterey County fruit.

Three years later, those same cousins decided to retire from the business, and I bought La Rochelle from them and moved it up to Livermore. At the same time, I decided—along with winemaker, Tom Stutz, that given the familial history with Pinot Noir, our relationship with great growers all over California, and the grape's growing popularity due in large part to the release of the movie, *Sideways*—to refashion the brand as one dedicated nearly exclusively to the noble Burgundian variety.

We were able to make some really delicious wines for the nine years the brand was in Livermore. In 2014, I wanted to devote all my attention to making wines from Bordeaux varieties, so I sold La Rochelle to Chuck Easley, a great friend of mine. His La Rochelle continues to prosper in Kenwood. o

CHAPTER
12

LIKE BIRDS SURPRISED INTO FLIGHT

Today is Thanksgiving. I put the turkey in the oven at the winery and walk hunched, pushing away the heavy cold, to our wine-making spot next door. It is dark out, and there is no traffic this early on Tesla Road; the stars hang thick on the black dome and are obscured by the steam coming from my mouth. This is also the last day of harvest.

I open the sliding gate with my magnetic card and walk through. There is one dimming bulb hanging under the eaves of the tent covering part of the small-lot winery's crushpad. The light is enclosed in a plastic cage twenty feet above the ground, and it makes shadows of the tanks at the margins of the tent. Most of the tanks are empty now because the larger winery we work with has already finished its season. Our fermentation regimen takes longer, and the last lot of Cabernet Sauvignon we pressed yesterday took longer to come together than most.

The cold is oppressive; the difference in temperature between the smooth-walled steel tanks and the air is negligible. I unlock the door to the 100 room, fight the absolute black of its unperceivable perimeter to a light switch, shed dim light, and put on a pot of coffee. It has been a good year so far. The quality of the wines is very high, even at this point, and the black emptiness of the last couple of seasons has filled-up some with motion and light. I leaf through the harvest book in the lab, waiting on the coffee burbling. We have harvested and crushed and fermented and pressed and barreled down 34 different wines this year. 35, the final one, is waiting outside under the tent. We have brought in fruit from six different vineyards, Cabernet Sauvignon from each of them. Multiple wines from some of them. The other varieties—Cab Franc, Merlot, Petit Verdot, Rhônes, and Italians from Ghielmetti Vineyard and the Home Ranch. I turn the pages and look at tonnage figures, dates for pressing, the numbers marking the fermenters, and it is hard, at the eve of the end, to remember all the thousands of man-hours over the last several months that have gone into creating this book of wine. The coffee pot gasps once more, I fill my cup, and head back outside.

Aidan and Beth are coming through the metal gate, and the sky has lightened when I leave the 100 room. They stand next to the last tank; I am distant enough so I can't hear what they are saying, but Beth is laughing and Aidan, tall and gaunter than usual, is talking to her animatedly. His fine hands move like birds surprised into flight. The white canvas tent is shiny on the outside and is tied down to the ground by guy wires and ropes strung through 55-gallon plastic drums filled with water. The inside of the tent is coarse, rent by wayward pallet jacks and smeared with the greasy graffiti of forklift collisions. It is still very cold, but there is little wind. In the afternoon, especially now, deep into November, the fabric walls beat under the wind like the full sails of ships. The flagpole to our east in the middle of the courtyard, now shining in a pearly light, seems like a maypole around which the gathering clouds are swinging.

"... these tanks?"

"We should rinse them." I can now hear Aidan responding to Beth's question.

"This is the last one?" I ask, gently knocking the sides of a 650-gallon tank to determine how far up the wine rests. A plastic placard looped with twine through two holes at the top is taped to the tank. The sign lists our bonded winery number and is supposed to be attached to all of our vessels in case the government comes auditing.

"I'm going to barrel it down. Beth is going to clean out the other ones."

"Champagne at 2. Remember," I say. Beth smiles. "Then we'll eat at 4," I finish.

"Okay. We should be done in a couple of hours. I'm going home first to take a shower," Aidan says.

"I'll be around for a while. I want to taste through a couple of early lots," I say.

Beth says, "Bring us something good to try."

"I'll do what I can," I tell her, smiling.

There are a couple of smaller lots of barrels stacked under the tent because this year was a big harvest and inside is full. I step high over a 2-inch hose that Aidan is dragging across the concrete. One end will be attached to the valve at the bottom of tank *S-740* (its identification painted black as a tattoo on three sides), the other to a curved metal tube that fits into the hole at the top of the barrel and fills it. The barrel lots outside are stacked neatly two-by-two in metal racks, three levels high. I look at the barrel ends and notice that just over half have that year burned into them below the woodcut Steven Kent Winery emblem. We've cut down on the percentage of new oak we use because we don't like the aroma or taste. The rest of the barrels are from two and four years ago. There are two lots, and each is a mirror of the other. We divide barrels up from five or six coopers so that we add organoleptic complexity as early as possible. No two Medium+ barrels are the same. Not even from the same cooperage from the same forest in the same year.

Quickly and unexpectedly, the flagpole is buried in mist; the canvas roof plays out a metronome tune, and the walls begin to snap. It has warmed up with the rain arriving. I take off my glasses, so I don't have to wipe them with my shirt and stuff them into my pocket. I turn back to look at the tent as I run to the 100-room door.

"Shit. Better move the barrels under the tent." I hear Aidan even as I smell wetted cement.

The rain is coming down hard, in bars, and through them I can discern Aidan and Beth moving disjointedly and quickly under the tent. Without my glasses, they look like the shapes of people on an old filmstrip just before it sticks and burns.

The timer goes off on my phone. The first hour for the turkey is done. Two more. I brined it in salt and sweet water with herbs overnight. It will be juicy even if I'm a little late getting back next door. The 100 room is cavernous again without all the boxes huffing away. We used 30 of them this year, a few, twice. LR-18 and LR-26 had cracked liners, so they stayed stacked up outside. Now, they are all washed and stacked. The whole lot will be brought to an off-site location in a month or so and stored there until we do this again next year. Next to the 100 room, through a passage, fifteen feet high, is the 200 room. This is the barrel-aging room. It is the same height as the fermentation room, but far bigger. We take up part of a corner of the space, and our barrels are stacked six high. When I was young, I used to love climbing to the top, one rack at a time, with a thief in my teeth and a wine glass in the web of thumb and pointer, to grab a sample. The rows were narrow enough that I could lean against one barrel and put my feet on the top barrel across from it, sipping. I wasn't afraid then, as I am now, of body meeting concrete floor, and OSHA made us stop the climbing anyway.

I am alone, and the 200 room is silent except for the muted rain. Even the bulk of this giant room cannot contain the reverberations of pump, and lift, and exhortation when wine is being made. The raucous guffaw explodes out through the swinging doors and lifted metal door to the outside to soak everything within the aural bloom. No less rich, no less tumescent now, in the quiet, a gentle but omnipresent infusion is the voluptuous conspicuousness of new wine aging in wood. I have a metal thief and a wine glass and ethanol in a spray bottle. I kick a small bucket over to a stack of new barrels and set myself up. The crew gave me the metal thief because I broke too many of the glass ones climbing up and down barrels. It works the same. I spray the thief inside and out with the alcohol to sanitize it and put it into the bottom barrel in a group of 12, two barrels slotted side-by-side on a rack, those racks stacked one on top of the other, six high. The tape on the metal band tells me the wine inside and the lot number. The tape this year is lavender. Every year is a different color so it's easy to tell one vintage from another.

The wine in my glass does not have any oak influence yet. The barrel is just the most convenient storage vessel. It will have wood soon, and the wood will color and enrich this embryonic fluid; the pure, pulsing fruit I'm tasting now. It is too early at the end of the beginning to make conclusions about the beginning of the end. The wine will go into bottle, will be birthed and will live its life nobly and completely if we have done our job right. It is not too early, perhaps, to contemplate how well this drowsing creature, gestating in its oval, wooden egg will greet that auspicious day. This is a time of innocence and hope. The French call it *elévage*—the raising up of the wine. My parental feelings for this year of wine and for my blood and flesh children know only a difference in degree. My four kids, seven years apart, all got a different father as he, himself, grew and changed; one that I hope was wiser and less prone to thoughtlessness as he aged and gained more wisdom. Each vintage gets a different father too. One, more sensitive and sure. Just as my children are my shields against the implacable advance of Time, carrying my name forward, my wines carry the family ambition and thrust the family shape—the line—ever outward.

I taste around from lot to lot, spitting into the small bucket at my feet. I don't take notes in my notebook as I would with older wines. Only first impressions, strange variances, interest me now. I am compelled to look for balance like I always do. The balance is glassy edged, and that is to be expected. Oxygen and barrel-time will sand down the sharp edges. I feel the buzz on my wrist again. One more hour gone; one more hour ahead. I dump my spit bucket into a drain, grab the alcohol bottle, thief, and glass and rinse them in the lab. Beth's looser hair is tightly cinched now, and there are spatters of wine around her tired eyes and on her cheeks. She tells me how many barrels we filled and how many smaller kegs we got from this last tank and leaves the lab more quietly than usual. I write the figures down in the harvest book. This vintage is very fine and will, with care, reward my team's hard work and my evolving vision.

I head for my truck through the 200 room and hear the huge swinging doors jam open at the end of the building. The storm is here full and rich, I think. It is Aidan, though, carrying racks of barrels from the last tank on his forklift. I didn't hear the rain stop, and the gap between the flapping doors show showers of sun outside. Aidan and Beth are exhausted. They've been working 12-hour days for weeks on end, and the more tired for seeing the end in sight. It is rare for us to be making wine this late into the season. It's happened a couple of times over twenty-five years, but it is the exception. Overlaying the rainwater, wetting the concrete of the crush pad, is a pond of wine laying its own viscous layer on top of Nature's. I see the filling wand left in the last barrel, and I silently confirm what Beth has already told me about volumes. And as I think about the process she followed to get wine from tank to barrel, and I look at the tank, now empty, and the valve at the bottom of the tank and the puddle of wine on the ground; I think I know what happened. Some of the tanks are old and they don't open and close as fluidly as they should, with one hand as the other grapples with the filling wand or a clamp or the miscellaneous other equipment that might be used. Exhaustion mixed with faulty equipment led to a mini eruption of wine from the valve.

"Beth?"

"Fucking valve," Aidan says, laconically, as he makes a new stack of barrels. The last of the year.

.... ii ....

I go back out of the sliding metal gate and walk back up Tesla Road to my place next door. The morning has come and passed and the sun shines rays straight down. The street is damp still, though only in patches. There are torn clouds way out west atop the sea-near ranges; the faint petrichor of wetted dirt hugs the ground. I have thirty minutes on my timer and I'm trying to organize in my mind the rest of the prep work. I'm a little clotted up from lack of sleep so my mental list will change.

I walk up the steps at the Tesla-side of the building, and wade through the smell of fire and meat to the kitchen. Thanksgiving is my favorite day. The wine and food of it, the blind-tastings, the laden table, family and friends unseen for a year are a great part of it; the planning, the stock-making, the shopping for it, the days in the kitchen, the taking care of my family, is the greater. I walk into the kitchen, and the heat of the stove in the corner is like a banked fire. There is a steel table in the middle loaded with plastic bags of vegetables and paper bags of flour and butter and containers of stock. I dropped this mound off early, before I went to the winery. I separate everything now, making smaller piles of the things that will go together.

In the refrigerator is a case of Champagne and miscellaneous bottles of white wine. I open a bottle of sparkling wine and pour myself a few ounces. I open the windows to the outside too. One year, we didn't open them, and the smoke from my mother's lobster skewers tripped the alarm. The firemen were roused from their firehouse meal and they were obligated to check for real fire. We have pictures of my smiling mother handing lobster skewers to the firemen in their truck as they were leaving. I take the turkey out of the oven, check it with a temperature gauge, and cover it with foil. I have a couple of hours before my arriving family washes up in tumult and joy and noise onto the shores of Tesla Road, so I walk down to the patio to enjoy my wine and the last quiet of the day.

The Champagne is good. It's a rosé from Bollinger, one of my favorite producers; it makes up in acid-driven fruit what it lacks in biscuity *sur-lie* notes. This style of wine would be easy to pair with food, but I'm just looking for a little revival right now. By itself, it is enough. The breeze picks up a little, and it breaches the house and eddies on top of the patio, I huddle down a bit. The browning canes of Sangiovese, just across the driveway, scratch against trellis wires. The canes still have leaves, though they are fewer than in summer and yellowed. Sangiovese is the first grape to bud out in the spring and the first to be harvested in late summer. There is an acre of it, and one-and-a-half of Barbera planted beside it. Small brands, unless they are anointed by the wine press and are self-capitalized, have little chance of world-wide success. If they are based in unfashionable appellations far from the remunerative power of Napa Valley, the difficulty is magnified. If they make great wine from the irrepressibly popular varieties such as Cabernet Sauvignon and Chardonnay, they can make a business for themselves. The larger part of the property, to the west of the house, is planted to the Concannon clone of Cabernet for this reason. There is work to be done now, in advance of the family's arrival, so I head back upstairs to the kitchen.

I get out a cutting board after pouring myself another glass of wine and peel carrots. I slice them on the bias and chop Italian parsley too. I'm going to parboil the orange disks briefly then sauté them in butter until they are just a touch firm. To the butter left over, I'll add the parsley and sea salt and use it for sauce. I set the carrots aside and begin cutting off the stems of Brussel sprouts. I love Brussel sprouts now. I didn't always. As I've gotten older, I've discovered the joys of bitter things. Perhaps they are like the wines I enjoy most, full of nearly scathing acidity. There is something cleansing and uncompromising about bitterness and acid. They cut through all the extraneous things.

The plan is to reduce the sprout to its individual leaves then fry those in the fat left over from the prosciutto that will accompany the green. We have a six-burner stove in the kitchen, a Ferrari engine in a Volkswagen. We were going to have dinners upstairs prepared by restaurant chefs that supported our wines. We'd do the dinners for our wine club members, personalizing the art of wine and food pairing. One of the other wineries in town, though, got hit by the Health Department for some cheese-related outbreak, and the regulations became much stricter after that. We didn't have the right sinks or drains in this 120-year-old house, nor the money to install them, so there are only a couple of nights a year when our stove gets to show what's she really made of.

I've got a couple of red peppers charring on one of the burners and the innards of the turkey—neck, heart and liver—boiling in a saucepan on another. I'm going to use the roasted peppers and tomato sauce, toasted almonds, day-old bread, olive oil, sherry vinegar, smoked paprika and salt to make a dip—Romesco—for quarter inch-thick slices of boiled red potato. I'm going to make an aioli too (egg yolks, lots of fresh garlic, olive oil, a touch of salt, pepper, drop of white wine vinegar). I'll chop up the heart and liver and use them for the gravy. There's music playing, Springsteen's *Darkness on the Edge of Town*, one of the top five albums of all-time, I've got my *mis-en-place* set up the way I like it, and I have a few more minutes before the house is filled with my family.

- 1 Turkey, brined. Liberally salt and pepper inside and out. Truss turkey.
- 1 bunch carrots, peeled, cut in half
- 2 fennel bulbs, washed and cut into strips
- 6 stalks celery, cut into 4-inch strips
- 2 medium yellow onions, sliced
- 6 parsnips, peeled, cut in half

- 1 quart chicken stock, freshly made
- 1 turkey liver, diced
- 1 turkey heart, diced
- Drippings from roasted turkey
- Smoked paprika, to taste
- Salt, to taste
- Pepper, to taste
- Dry white wine, 1 bottle
- Cream, 8 ounces
- Butter, 3 tablespoons

I am lifting the warm turkey out of its pan when I see through the foggy window my cousin and his wife pull into the parking lot. I put the turkey on the cutting board, grab a couple more wine glasses and pour Champagne into each. The upstairs door opens, and Mike and Barb come into the kitchen with their arms full.

"Wine flows like glue around here!" Mike says as he lays his load on the perimeter counterspace.

"Never gets old," I say, handing them each a glass of wine.

Barbara is from Switzerland and retains some of the reticent, Germanic hardness in her speech. She met my cousin when she was an *au pair* in San Jose, just out of college, thirty years before. She is tall and very beautiful. Much more forceful, I have learned over the years, than her demure manner and pitched sing-song cadence would lead you to believe. They have two kids, but both are away this year. All of us go outside to get the last of the stuff that they brought.

- Spread carrots, celery, fennel bulb, onions, parsnips evenly on bottom of roasting pan.
- Place trussed turkey on vegetable raft. Leave turkey undressed to aid in drying of skin.
- Pour half bottle dry, white wine on vegetable raft. Reserve the rest.
- Roast turkey at 425° for 3 hours, checking each hour with thermometer.

- When turkey reaches internal temperature of 165° and skin is brown, it is done

The elevator just outside the kitchen whirs and a few seconds later, Nancy and our three dogs erupt from the small cage. She has her hands full of wine and platters and bags. Barbara and I help her unload. Mike pours her some Champagne. We kiss, and the four of us touch glasses. The three dogs exuberantly sniff for dropped food then range quickly from one room to the next. Their nails clack out canine rhythms on the wood floors.

My father and mother arrive up the steps, weighted down with glass trays and wine. Their two dogs add their considerable energy to the environment, a few more puffs of magical air into the expanding balloon. Mike opens a bottle of 1990 *Pol Roger* Brut Chardonnay. Mike knows, because we shared a bottle together many years before, that that wine was the best Champagne I've ever had.

In reality, whether the 1990 *Pol Roger* is the best Champagne I've ever had is to miss one of the fundamental truths about wine. The *quality* of a wine, how good it is, is inextricably tied to the experience in which it is drunk. No matter how objectively good (how well made) it seems, the wine will be colored by experiences profound and debauched. I've learned that it is easy for individual bottles to become commoditized in my mind, institutionalized by profligate access and the unexceptional rolling out of days. The wine bottle deserves a greater and more individual memorial, and if allowed to, will usually make the prosaic day something worth remembering.

The day we drank that wine together, Mike had come over to my house to collect his son, David, who had spent the night with his cousins. Our kids were making short movies in the backyard, they took turns being serial killers and corpses. The weather was gorgeous, and Mike and I were talking about our mutual musical obsession, Bruce Springsteen, on the patio outside. We'd take turns choosing songs to play, opening up more wine. I had a wipe-off board in the kitchen, and we each attempted, over the next several hours, to list our *Top 10 Springsteen Songs of All-Time*. I woke up late the next morning, got coffee, and noticed the board on the wall. I had forgotten our lists. The blood-splatter patterns on a wall will tell a detective the story of the circumstances of violence and death, the legibility of our *Bruuuuuuuce* list told a tale of the murder of sobriety. Each successive song title was less legible, and one *Top 10* list had nine songs, the other, eleven. *Pol Roger* will always be *that* night, indelible and gorgeous.

Aidan and his boyfriend arrive in the parking lot outside, and his oldest sister, April, with her family, pull in a minute later. The kitchen is getting warmer with all the bodies. The new arrivals add their dishes to the shrinking counter space, and Nancy leads everyone outside to the patio for a glass of wine. We wait for the elevator down, bottles in hand. The elevator door opens, and one of my dogs, peopleless, is sitting on her haunches, alone, in the middle of the lift. I turn to Nancy as Tess pads out to the kitchen, and I shrug my shoulders, smiling.

Mike, the brother I never had, always brings the cheese tray: Morbier, Mimolette, blues, St. André, almonds, crackers, Brillat-Savarin (the greatest butter on the planet!), olives. Sara and Katherine, my two youngest, who I missed arriving, are digging in. I bring more chairs from other tables to the main one and sit down for a moment. I have been hosting Thanksgiving at the winery to a revolving—contracting and expanding—guestlist for more than a decade. In all these years, on only a couple of occasions, have we been unable to start our day here, outside. California is expensive, but outside with family and a glass of wine at the end of November pays the freight every time. My father opens a White Burgundy from *Grivault*. It has some age on it. I look through the glass, and the vineyard behind is gilded. More family and friends arrive, and the moment keeps expanding.

Before we head back to the house, my stepmother cuts some canes (those on the Cabernet side still have green leaves) and collects a few pods from a magnolia tree. She is good at decorating the tables. We grab glasses, bottles, and trays and walk back down the path to the house. As I enter the cellar, I hear the dogs, now six of them, scrabbling houseward through the vineyard, racing to follow.

- Place the vegetables, offal, paprika, salt, and pepper into a blender.
- Add the liquid from roasting pan, stock, and wine and blend until desired consistency is reached.

In a proper house, the kitchen is the cultural and physical center. Right now, there are 10 people crowded in, plus the oldest and smallest dog, Capi, several open bottles of wine, all six stove burners ignited. I'm peeling potatoes for garlic-mash, 10 pounds. I have a glass full of Chinon from Joguet and am loving the fruit. I finish peeling and I slice the potatoes into small pieces, so they cook faster. I raise my head and wipe the sweat away with a towel. The ten are broken up into small groups that keep morphing as individuals disengage and move from spot to spot. Folks are reaching for appetizers over there, readying their dishes for the oven over here. There is chaos, but it is cordial and slow moving. We don't see each other often during the year, and it is as if we are fixing in our minds what the others look like now, comparing each to a picture we had in our mind from the last time. Which of us have gotten older and fatter? Is she ok? She looks tired. Mostly, we connect, disconnect, and reconnect in these small groups to feel, again, the peculiar warmth of the ones we love the most.

- Pass the liquid through a fine-mess strainer into a saucepan on medium heat.
- Add the cream and butter with a whisk. Add stock if too thick. Add salt and pepper to taste.
- Warm gravy through and pour in gravy boat for service.

Dishes go into and come out of the oven, the turkey is carved, and the food is brought down to the cellar for dinner.

I'm the last to get downstairs. I carry some glasses and some nearly empty bottles with me. Aidan and Beth are opening bottles of wine that they brought to dinner. My father is tasting through them to make sure they aren't corked. Sleeping dogs litter the concrete floor, and the line is forming at the buffet. I'm tired and a little drunk. Thanksgiving starts at the beginning of the week for me. I shop on Monday, make stock and brine the turkey on Tuesday, do the prep work for my dishes on Wednesday, and cook on Thursday. We are still working too. I'm tired but this is the way you take care of people. You nourish them and bring them close to you with food and wine and love. Our gatherings have included friends who are no longer with us; babies who have grown around this table and now have babies of their own; friends who have left my employ but are still here. Once you are invited, you are invited for life. My late wife, June, shared this vision, though she knew this was my day and took care of the front-of-the-house (Christmas was hers!). Nancy, too, is also deeply connected to the caregiving aspect of this day. She is still working her way through the family dynamics.

Plenty of wine, mostly mine, is drunk through the meal; each of us, in turn, tells the table what we are thankful for (my youngest, Sara, when she was 10, told the assembled group that she was thankful for her 30-year-old boyfriend that she met on the Internet. The delivery was perfect and priceless, as was the horrified expression on her grandmother's face), dessert is had, dishes are done, wine bottles are inventoried, music is turned off, and late in the night, homeward bound we are.

I get back to the winery in the morning and grab myself a cup of coffee in the kitchen. I am alone today as the rest of my team is enjoying a well-deserved weekend. Everything is quiet. I smell the turkey still and the bouquet of Cabernet Franc as it explodes out of last night's glass, and the subtle scent of Barbara's perfume when we hugged in the lee of yesterday. I taste the gravy from my finger, taste the taste of Nancy when I kissed her. My family and friends linger here in today's empty and silent kitchen, incorporeal reminders of who is most valuable to me and what it means to work for the joy and richness we have the privilege of adding to other people's lives. ○

CHAPTER
13

IMPRISONED IN THESE BARRELS

From this aerie upon the last rung of the tallest ladder, I can see the many sweat-sodden days and sunburnt days of the harvest. The dark mornings are imprisoned in these barrels; the 18-hour days and the muscle-stretching work, the thousand small cuts, the orgiastic smells and flavors, the presaging of births and marriages and memorials; of the sun-drenched summer day; of the wind-stirred, fog-bound fall mornings—all are caught up and held within the confines of this curved and wooden space. Held unmoving in its gentle forest arc is this magical drink, thick with life, indeed, transmitting its reverberate beat parallel to the tick of the clock and right-angled to un-parenthesied space. Take away the freighted subtext—the *meaning*—and barrels are about the perfect vehicle for transforming a liquid, all corners and straight lines, rays and chords (especially when young), and making of it something round and deep and connected. The barrel draws the circumference around the wines we end up drinking.

The wood we use for barrels is the white oak species *Quercus alba*. It has the wonderful ability to hold liquid tight yet also allow air to move through its microscopic pathways to micro-oxidize the wine inside. Oxygen molecules move freely in and out of the oak enclosure, evaporating alcohol on the egress, lengthening tannin chains and rounding off the wine on the ingress. Barrels have been used since before the birth of Christ to hold wine, giving over to it not only an oxidative effect that sands off the sharp corners of acid and tannin but also infuses the liquid with flavors and aromas ranging from butterscotch to burnt coffee. Coopers, those trained in the art of barrel making, toast barrels over an open fire fed by the same oak—scraps that are left from the hewing out of staves. The cooper will take the barrel staves, joined together at the bottom by a stainless-steel hoop and splayed out the rest of their length (like fingers on a stretched hand) and set the incipient and open barrel over the flame for a specified period of time. Together, the temperature of the flame and the duration in contact with it will lead to a new barrel caramelized to a foreordained set of flavors and aromas and to pliable wooden fingers that are bent by machine and engirdled by additional hoops to the final, familiar shape.

Each cooperage has its own recipe for toasting. Most of the barrels we use, made by a half-dozen artisans, are *medium toast*. This toast level in *François Frères* is quite different than that of *Taransaud*. In addition to the proprietary nature of the toasting is the choice of forest that the trees come from in France. Most of the wood we use comes from the Center of France, quite northerly, in the middle of the country. Oak trees for barrels flourish in cool climates where their rings grow more tightly together, providing for a tighter-sealing barrel as well as for a more modulated giving-over of the organoleptic (flavor and aroma) components and structural elements we pay so much money for. Many wineries use oak from trees grown in America too. The Midwest—Kentucky, Missouri, and the forests growing verdurous along the spine of the Appalachians in Virginia—provide the best of American wood. Not liking the overt flavors and aromas of toasted oak that interfere with the beautiful nuances of well-grown fruit, I choose barrels that impact my wine structurally but show little in the way of spice and caramelized sugars. American oak is far more impactful and less elegant than French; the wines we made from American oak often seemed to have subtle scents of dill or coconut, neither of which made the wine better. In 2009 we stopped using American barrels and began aging our wines in French oak exclusively.

The proper use of barrels is a hallmark of good and thoughtful winemaking. Wine should be, above all else, beautiful. The virtuous relationship between fruit and acidity, tannin and wood, is what leads to balance. And balance leads to Beauty. Too much new oak with all its primary puffing out of chest leads to a wine with too little fruit and, by default, too little Beauty. Not enough new wood or a wine in barrel for too little time leads to thinness and lack of texture—unbalanced, and unbeautiful. Sometimes the proper use of barrels means no use of oak at all. Each wine has an essential *it*ness. It is as if there is a vinous shape wafting above the winery for each load of fruit that comes onto the crushpad. Inherently, this shape is perfect—it is the absolute culmination of that block of Cabernet Sauvignon or Pinot Noir. It becomes the winemaker's job, then, to identify this spectral doppelganger floating over the tank farm and bind it together in harmony and balance with its pre-wine self, incarnating juice into its inevitable alcoholic twin. With certain of these spiritous spirits, the eschewment of oak is the only possible right choice. The fully-fledged, acid-washed, fruit-engorged wine becomes the honest and good one.

The solitude of my little church is what I crave the most. My Friday sanctuary, *l'Église de St. Vendredi*, helps me ratchet down the noise of the rest of the week, of the sales, the search for investors, and the putting out of many small administrative blazes. On Fridays, those practical considerations kindly disappear behind an esthetic, nearly hedonistic, curtain that is the winemaker's kinship with his wine.

Because wines change so dramatically during the course of their lives, it is necessary to taste them frequently to see how they are progressing. Typically, I will choose a lot or two of barrels, and armed with chalk, alcohol, a thief, and a glass, I'll taste one barrel after another until I've made my way through them. As I'm sipping, I'm thinking about how the wines smell and taste, how much wood they have picked up, how they feel. Each sample gets a nosing; I will put my hand over the glass, shake vigorously, and put the rim up to my nose. Because it is cold here, aromas and flavors tend to be a touch muted. Experience helps you to project out what these wines will ultimately taste like, but the shaking helps to get the wines to open a bit to make that supposition a little more concrete. After smelling, I take a sip and swish the wine around my mouth energetically. Again, I am looking to get the wine moving and opening and hitting all parts of my mouth. I'm drawing a lot of air in at this point too. As I spit the wine out into a bucket, I'm sensing the length of the wine, how long the flavors persist, where the tannin hits my mouth, on my cheeks, along the sides of my tongue? I'm then marking the barrel with a grade so that I can later see a running record of how that particular barrel fared against the others in the same lot and how much it has changed over time. I am also thinking about, even this early in the process, where this barrel might ultimately fit. Is it destined for Lineage—our finest wine—or for a humbler offering? I have found over the years that these early observations are often prone to significant change when the final blends are made. Barrels that I gave a ++ (my highest grade) to early on do not always make the final cut for the best wines. Even individual lots of wine, a single clone of Cabernet from our Ghielmetti Vineyard or the best Cab Franc from Sachau Vineyard, which seemed so promising at first, will get relegated. And less good lots often evolve so beautifully during their time in barrel that they must be used for the best. This propensity for indeterminateness can be daunting; once the final blend is made, you cannot unmake it. But when you understand that these wines have a way of finding their true selves, barrel by barrel, and that it is my job to facilitate this meeting, the daunting becomes the magical.

The plastic bottle that I mentioned before is 70% alcohol. We spray our thief (a glass tube about a foot long that we insert into the barrel to draw a wine sample) with this solution because it kills practically anything that might cause a problem in the barrel. This simple prophylactic is the difference between a wine that shows a purity of fruit that makes the bishop weep to one that tastes and smells as if a bunch of sweaty horse saddles had been tossed into a fermenter. Like all living things, wines can be afflicted by any number of "diseases" while they're aging away. The spoilage yeast, *Brettanomyces*, Brett for short, is one of the most prominent of the bad bugs that ubiquitously inhabit the winery. It is this same yeast that is increasingly difficult to kill off and that gives off those immediately recognizable sweaty-horse/latex-band-aid aromas and flavors. I marvel at the documentaries of French winemakers sampling their barrels one after the other with no sterilization of the thief between them. This situation is like STD-hell. The ease at which one infected barrel infects another is akin to the dude with syphilis run amok in a warehouse full of virgins. In California, Brett is considered a winemaking flaw, one that the judicious use of sulfur and proper sanitation can fairly readily combat. The challenge with Brett is that it is extremely difficult to balance its very obvious organoleptic contributions against the more subtle contributions of fruit, fermentation, oak, and aging. Europeans, though, have a different view, I think. Their wines have contained Brett for so long that its contributions have become part of the cultural legacy of the producer and the place. "Clean" wines, like those favored in the New World, would be as foreign to the drinkers of Burgundy and Côte Rotie as their odiferous wines are to the typical wine drinker in California. I don't fall as heavily in the "clean" camp as some of my fellow winemakers, as I think a touch of Brett adds complexity to a wine. And interesting wines are always worth tasting.

Noticeable Brett or overly high Volatile Acidity (VA), another flaw in the New World and one fairly easily guarded against, have, of late, been seen as strange badges of honor in some of the "natural" wine world. I imagine these winemakers as priests in some abbey, miles and centuries away from any center of habitation, consulting esoteric texts, devoted to a system of winemaking that is "pure" and intransigent. There is something admirable about devoting oneself to a cause, unconcerned with commerce, making a style of wine that hews as closely to one's precepts as possible and is not necessarily influenced by what is happening in the outer world. I wholeheartedly subscribe to this notion too. Where I diverge from the "natural" winemaker is in my relationship to the consumer. I'm not a hobbyist, nor am I making wine only for my family or friends. I have a responsibility to provide my customers with a wine that is compelling, ageworthy, delicious, AND stable.

I do not have the luxury of providing my customers with experimental wines that have failed the fundamental test of providing real value for the dollar spent. I have made a bunch of wine over my career that didn't reach a level of quality that I would want to share with my customers, let alone charge them for, and these wines never crossed my customers' lips. I will fail on my own dime; I will not ask them to underwrite my own enological ramblings. This is not to say that I won't offer my supporters some interesting, even weird blends they haven't had before, because I will. The appreciation of wine is mostly about being willing to expand one's sense of what is delicious and right and to have a curator, like me, to help along on the journey from the most comfortable and familiar to that which blows off the socks and drops the panties. o

CHAPTER 14

PERFECTION IS THE ONLY THING LEFT

What makes a great wine, in my opinion, is a sense of inevitability. Great wines are or become what they are meant to be. They have a sense of cohesion and a sense of propriety and a sense of promise. They do not show all that they will be in every sip; instead, they show you balance now and a hint at their future and full glory. Great wines are always in the process of becoming—perhaps an apotheosis in the greatest of examples—but always on their way to fulfilling their true natures. Balance is certainly a great part of it. A great wine by definition is one that on its best day is utterly whole. There isn't anything more or less you'd ask of it. That wine has the right combination of weight and fruit and acid and tannin and length. All the corners have been sanded and you are left with a perfect sphere. Or to put it another way, like Michelangelo's David, all the excess rock is broken off until perfection is the only thing left.

Unlike that statue, though, the state of perfection is unbearably transitory. And it is this quality, this briefness, this mutability, this only-of-the-moment*ness* that is the true glory of wine. Really good wine gives a pleasure that is thrilling in its physical briefness, and very long in the memories that it creates. A great wine has the power to transfigure Time.

If you read the major wine magazines such as the *Wine Spectator* or *Wine Enthusiast*, most of their reviews of high-end wines are accompanied by a recommended drinking window. This "massive" Cabernet will last for 20 years; or drink from 2020–2030, etc. Research shows that more than 90% of wines that are bought in supermarkets and wine shops are consumed the same day, so the drinking window (in this context) is essentially meaningless. But that's a topic for another rumination. What's not meaningless, however, is the notion that part of a wine's essential quality is tied to the perception of how long that wine can age or how long past release the wine hits its peak. This equation is influenced by all kinds of caveats, and I have spent a career thinking about its validity.

America is a very young wine-drinking culture, and California a very young wine-making area. When you are just beginning to become interested in something you tend to look to examples from the Old World, no matter how dissimilar the growing areas are. The notion of ageworthiness as a prerequisite for quality was formed long ago and based on a foreign paradigm that doesn't seem relevant to me anymore. If one's model wine is hard and tannic and needs decades before it is drinkable, and that need for extended age is viewed as a *sine qua non* characteristic of world-class quality, then the new winemaker needs to figure out a way to induce this kind of structure in her wines. It may not have been clear in the 1970s that we can only truly produce California wines in California; the grapes and the sensibilities of California winemakers may be the same as those in Bordeaux, but no terroir can be veritably translocated. Our wines are what they are, California will never be Bordeaux, though with global climate change and the pre-eminence of certain wine critics, the wines of Bordeaux grow ever closer to California in style. We may soon see that the obvious differences in Old World and New (especially in recently released offerings) are smudged into immateriality.

When my father really got into wine in the mid-1960s, the best examples of Cabernet were Bordelais and these massively tannic and acidic wines needed decades to soften and complex up. The same was rarely true of the California example of the grape—especially those wines made in the last 10–15 years when the predominant wine-making style shifted to longer hangtimes, more fruit, less acid, and less ageworthiness.

One of the great truths and part of the essential nature of the greatness of wine as a thing is that it does CHANGE. Wine evolves; it matures; it lives out its life sometimes grandly and over a great many years and sometimes its nature is that of the Mayfly. How wine changes, in a practical sense, is for a long chemistry conversation. But HOW a wine is going to change, or AGE is fraught with so many tangential but crucial questions I'd argue that it is knowable only within a range, and not even very well at that. Storage conditions—temperature (absolute temperature and temperature range), humidity, absence of vibration and light; quality and integrity of the cork; wine variety; reputation of the producer; winemaking style; appellation; hillside or valley floor... all of these factors will influence the wine's ability to age. Then when you take into account your personal preferences (freshness of fruit, like or dislike of tannin and acid and wood, desire for tertiary aromas and flavors brought on by bottle bouquet and age) age-worthiness as a signifier of quality becomes much more complicated and ephemeral a conceit.

The wine press and the avid wine consumer have taken a simple Old World reality that high-quality wines need time to become enjoyable to drink and have done their transmogrifying voodoo on it. What we are left with is a tarted-up New World notion that ascribes excellence to a dynamic that began as a purely pragmatic realization. I contend that there are no objective measures where affairs of the heart and the palate are concerned. I can measure the amount of titratable acid there is in a wine, the pH level, the amount of alcohol by volume, and a bunch of other things. What I can't measure, though, is how these things make you feel, how all of these individual planks combine in your mouth and your mind and your heart to create your individual sense of the *wholeness* of the thing. To *know* the details of a thing can be revelatory, but the details rarely capture the essence of it. The esthetic side of life (the wine side of life) is about creating emotional connections, about intuition. To know something on the wine side is to suppose and to assume and to guess. Facts are for another view of things; a view of life that does not welcome the *heart*, for me, is a view that is incompatible with everything for which wine is a symbol.

All wines can age. Not all, in fact, —most—are meant to age at all. Each time you drink a wine, you get only a snapshot of its life; the quality of that life, however, is purely a product of your own needs and desires. Describing the color orange, expressing in words the essence of the thing you are tasting and smelling, is nearly impossible. Great wines tend to get more complex the older they become. Primary fruit and the impolite, exuberant structure of youth give way to integration and propriety and the depth we associate with maturity. Neither state is inherently more valuable or more worthy than the other. These states are not even individual stops reached in the whole span of life; they are only moments in time, fated to change even just a moment later. Really good wine gives a pleasure that is thrilling in its physical briefness, and very long in the memories that it creates. A great wine has less to do with how long it can live and more to do with its ability to transfigure the Time that it has.

.... ii ....

In 2019 we celebrated the tenth anniversary of our Lineage Wine Company brand. I had the opportunity to lay all 10 vintages out in front of a wine writer one day at the winery. Because we are always so focused on the present and future in winemaking, I had not had the chance before then to taste all ten wines side-by-side. I was taken aback by how young the oldest wines still tasted and felt at the same time how their complexity grew even in the short time the wines were in front of me. Each of these wines not only captured a time and a place but also a changing palate. I may be a lot of things, but resistant to change and reevaluation in the face of a better way of doing things is not one of them. The older I have gotten the more I have prized acidity, pace, and freshness in my wines. Balanced acidity, to me, is the single most important answer to the question: what makes a wine worth drinking in the first place? Without the litheness and aliveness and pace and sexiness of a beautifully acid-driven wine, you are left with flabbiness, too little verve, and a wine that has grown moribund and lazy. This flight of ten wines showed a lot of things done really well, and enough work in front of us to keep us engaged and focused for the next many generations in our quest to produce one of the finest wines in the world.

That day in 2019 was an important stop. For that couple of hours in which I revisited the past through the acid and tannin and age and development of my wines, I had the opportunity to recollect on all the impulses and thoughts jangling around in my brain that fueled the final decisions that resulted in those wines. Each of them embodied the sum total of sun and wind and experience and attitude and hope and mission up to that time. As I tasted 2007, the first vintage, I remembered looking at that perfect fruit coming across the crush pad and thinking that everything that came to the winery in the years proceeding that moment was but a prelude to the creation of a wine far different, more focused, and more ambitious than any of the hundreds that I had made before. 2014 was the vintage of my wife's death sentence; 2016 the vintage of that sentence carried out. Each of those wines, those first ten wines of Lineage's history are so meaningful to me, I cannot be unmoved by any one of them. If one works purposefully to make of his work life something that represents the finest in him, the most meaningful thing about him, a symbol of his life…those wines are, for me, that symbol. It is difficult communicating in this way; somewhat embarrassing, perhaps. Earnestness is not a quality much prized these days. In this time of *fake* everything, it can be counterproductive to be nakedly real.

While I will always be my greatest critic and will find the unforgiveable flaws in my wine when no one else can sense them, I do desire to share them, nonetheless. In all their stages of growth my wines are honest representations of all I had to give to that moment, and welcome or not, they are the only tangible gifts I have to give. The gifts are given in good faith and cannot be ungiven no matter with what indifference they may be received. Being as much a whore as the next businessperson, I was hoping that the writer, who had never visited Livermore before nor had written a word—positive or negative—about the wines coming from there up to this point, would choose to make Lineage one of her subjects. Such was not the case, however. I'll take it as a failure on my part that I was not sufficiently compelling to convince her of the worthiness of Lineage as subject. It certainly was not the wines. We may forever wonder if the tree falling in an unoccupied forest makes a sound; I am confident, though, that despite the silence from this writer, those wines sing sweetly and loudly to those attuned to the right frequencies.

............iii............

I was looking out the window the other day at our vineyards growing lushly in the late Spring. The vines were trellised, and the vineyard was uniform, and the riotous odor of bloom was carried on the breezes. We work hard to make sure our vineyards provide the best fruit possible for as long as possible, and their planting was well-conceived and well-executed about 15 years ago. This manicured imagining is duplicated all over wine country and the fruit from these many thousands of acres of vines is worth billions to their owners. A day or two later as I was walking up to my office, I noticed a rogue vine creeping along through the landscaping. It was probably the product of a seed that had passed out the ass of a Starling and it was doing what its many millions of grandfathers had done over millions of years, spreading copies of itself across space.

What struck me in the juxtaposition of the vineyard and the vine is the constantly evolving attitudes of man to nature. Hawthorne saw chaos and uncivility and Godlessness in the forests of Massachusetts. Contemporaneously, Thoreau saw the immanence of the Creator in the very same woods; his Walden Pond relationship was a reconciliation of man to the original Garden. Today there is a healthy debate going on between "conventional" growers and those "natural" winemakers and grape growers who preach (sometimes with a shrill timbre) a minimalist philosophy about their craft. These latter would contend that the natural energies of a vineyard site become misaligned and depredated by the use of chemicals to ward off infestations of malign insects, mold, and fungi. Perhaps rightly, they believe that the farmer's vigilance in the vines is the greatest prophylactic and obviates the need for chemicals. These biodynamists are convinced that the addition of esoteric teas derived from natural products and applied at times corresponding to the cycles of the moon are the only sprays a vineyard needs to produce healthy vines and grapes and wines of unique energy and liveliness. There's no real way to tell whether these wines, which are then generally made without or with very little sulfur, are in any way superior to wines made from conventionally grown fruit. The religious implications inherent in the *natural wine* philosophy along with the purely subjective and individualistic nature of tasting and evaluating preclude a non-subjective, non-emotional conclusion.

Despite the lightest touch of man in the vineyard and the cellar, the vineyard and the wine are man-made objects. The vine I saw creeping along the ground searching for some vertical guide to bring it sunward has no chance to produce fruit that will be made into wine. Modern wine is a product of rational thought, experience, specialized equipment, chemistry; the vine is pruned and leaf-pulled, and shoot-thinned, hedged and sprayed, harvested by hand and machine... its genetic gift is wrapped up in a quilt of seed and skin and juice that obtains not only to the promulgation of a species but also to the physical need of the ones (we, who enthusiastically drink of its gorgeous nectar) who guarantee its spread. Its biological imperative is fitted up quite nicely with the higher-level desires of the human such that it innocently and effectively hijacks our enthusiasms by contributing to a virtuous circle of gentle mind-alteration in exchange for propagation and protection.

There's a wonderful scene in the movie *Jurassic Park* when the Jeff Goldblum character, weary of the hubris shown by the scientists who switch off the sex of some of the dinosaurs they have created so only one gender remains, says that no matter how advanced the science, no matter how thoroughly these scientists have assumed the role of God, nature finds a way to endure. It was true in the movie, but not true in the vineyard. No number of monkeys banging on keyboards can ever create a First Growth. That final product is a derivation of human desire and human intent. Ultimately, the natural world deserves our concern and protection. The vineyard and the winery, though, are not truly part of that world. No matter how careful the farmer and non-interventionist the winemaker, the relationship of the man to the material is a benignly exploitative one. ○

"When I read Faulkner or look out the windows of Matisse or thrill to the textures of a great Bordeaux, I am headed on a journey that is the same as the one that I hope my wines encourage."

CHAPTER 15

A CLEAR AND MELODIOUS SONG

I'm sitting in my tasting room, just after harvest, with a wine writer tasting and talking about the current releases of our wines. Along with the new wines, I have added a couple of Cabernets that are 10 years old. I want the writer to be able to taste and sense how our wines change, how they become more and more what they are supposed to be with the turning over of the calendar. The longer we talk, the more disappointed I get. He doesn't get it. He is so focused on the minutiae—barrel type, fermentation regime, the purely technical tools brought to bear on the making of the wine—that he is blinded to the crucial emotional impact the wines make—the way they make you feel, the boulevards of vinous and personal potential they reveal. I realize that we, the writer and I, can never be compatible, that our vision of the world is too dissimilar.

When I read Faulkner or look out the windows of Matisse or thrill to the textures of a great Bordeaux, I am headed on a journey that is the same as the one that I hope my wines encourage. More than the obvious perception of the words on the page, colors on the canvas, or aromas of fermented grapes, this journey is fundamentally emotional. Experiencing great beauty forces one inward to an interiority that reveals new possibilities and truths. It allows one to overcome (if only briefly) the hegemony of habituation and rewards the trying on of a new skin. What one thinks in this new place is far less powerful than what one senses; intuition reigns here in this inner place, and where there is intuition, there can be magic.

Perhaps this sense of intuitive connection is only another concept rightly called *epiphany*, and I have had an epiphanic response to wines a few times in my life. A year or two after we started Steven Kent, my father gave me a bottle of sherry from fruit grown in the Central Valley. His father had given it to him 25 years before, and he thought the wine was probably about 70 years old when I took possession of it. I knew only the smallest bit about fine wine, reading up in the wine magazines about those wines that were regarded as the *sine qua non* of their type: first growth Bordeaux, DRC, Gaja or Bruno Mascarello from Barolo; Napa.

I knew enough to know that Rancho Philo sherry from Cucamonga was no one's prize. It might be interesting I remember thinking then. I opened that bottle on one quiet Christmas celebration 70 years after the wine was made and took a sip before I passed the bottle around to a small cadre of employees. Something changed in me after I tasted that wine. That humble bottle carried its age with grace, and it sung sweetly of years of plenty and years bereaved, hauled along those aromas of vanilla sugar and whiskey barrel, of stropped leather and raisin, and filled the mouth with a sweetness and a viscosity that I can feel now coating my tongue even 25 years later. I do not know who made the wine or if that company is still transmuting fruit into such powerful memories; but that wine was the first hand grenade to blow apart my misconceptions.

The wine told me everything I needed to know about that time three-quarters of a century ago, about the weather and about technique; it told me also, and more importantly, about the people who made that wine; those who picked the fruit and crushed it and fermented and aged it, who added alcohol to it to make it strong and durable. I did smell and taste the tribulations that affect all makers of wine in that luscious dram, and our true communion, one winemaker to another, one creator to another, was experienced most truly in a place on the other side of the rational.

I think my best wines touch sympathetic drinkers the same way. The 2017 vintage, wines being released just as these words are being written, represent the finest quality that I have yet achieved. The terroir-driven details of that year roll out as perfectly as our understanding of the infinite complications of nature allow. The drought-dried soils were wetted prodigiously by winter rains in January, the dissolved solids of years of salted waters were washed from the root zones. The vines took up their nutrients from the unsoiled soil and grew without barrier in the temperate and long days of August and September. Our farmers removed the right number of leaves, laying the east face open to the beneficent morning, and the slow-to-ripen bunches were fed back to the earth, green, to leave finite energy to those berries hastening to purple. We made the right calls on the pick, the separate ripening threads of seed and skin and sugar, acid, and tannin unfrayed and wefted tight. The fruit arrived at the winery cold and early and went through the machines that pull bunches apart in the fashion they were meant to. We employed the correct fermenters for the right fruit; gentle and ambient for the fragile berries; hot and extractive for those whose structural glory shows best under duress. We pressed at the right time, chose the right barrels, the right time in, and the right time out. We spent the right amount of time blending, choosing the constituent barrels correctly, and making no more than the 80th mock which happened to be the best of them. I did not know for sure at any one step that we had reached a 165-year culmination; that would have to wait for a more introspective moment. I knew though, at each step, we were working with something grand.

The finest wine of that very fine vintage is Lineage. It is different than any of the 10 wines that came before it in terms of the final blend, and different, more crucially, with respect to its completeness and nearness in form and organoleptic complexity and overall weight to the picture that I carried in my mind of what we were at some point capable of producing. Much of what makes this wine great was achieved in the flesh and cell of that perfect growing year, a product of man's luck and nature's consummate and vast capacity. It is not a perfect wine, as no man's effort can be, and it will not be the best wine we will ever make. But it is a clear and potent stop along the pathway that marks progress and understanding.

This wine sings a clear and melodious song now, and its voice will grow in depth and timbre; its structure will hit notes baser just as its fruit will lithely dance along the treble clef. Nothing marks wine's uncanny nature more indelibly than its compulsion to evolve. Each sip is a life, an unfolding perhaps, of shining wings; perhaps, too, a long tumble into oblivion. Down its long line this Lineage will whisper stories to those who drink it just as that 70-year-old sherry did to me. And for those that venture to that inner place, as I do, the words of this wine will vibrate out as rings from their source, ever and concentric, to lap up upon receptive shores. o

CHAPTER 16

HEIGHTENED MAGIC

I never met a man truer to himself than my buddy, Claude Bobba. If true to yourself means facing the world with the face that god gave you, that is Claude. There is nothing deceitful about his demeanor, no saying this but meaning that. He is a true-blue guy. I met Claude back in the early 2000s and he was instrumental in helping to make early vintages of Steven Kent wine and for teaching me much that has helped propel our brand forward. Another friend, Mark Clarin, who worked as a winemaker for Wente in the mid-1990s, was the first to assist in the production of our wines, and Tom Stutz, the winemaker first of Mirassou then of La Rochelle, taught me much about vineyards and the right way to make Chardonnay.

Every winemaker but the first has learned from someone else and has taken that knowledge and experience and expanded upon it, if he or she is worth anything, and made it her own and then sent it off into the world. And the next one has come along and used the best parts of that gift and discarded the pieces, like the bits of orange rind, that don't get that one closer to the winemaking truth. Winemaking progress is a function of working through the techniques that the teacher holds dear, filtering them through your native intuition, tasting what those techniques and philosophies manifest and, ultimately, proceeding or diverging. The teacher is the trunk of the winemaking tree and the descendants are the branches, or so I thought. In reality, my teacher was one branch, and he came from another branch (there being no true trunk, as that person is lost to a time that no one can fathom), and I come from his branch, to branch out and to bud from mine, my son and Beth and those (hopefully) myriad others that will take forward the glories of the Livermore Valley and a germ of the truth of how to make the most from the fine vineyards of our Valley.

What I learned from Claude was technique driven. Claude was a home winemaker and did not have the advantages and the disadvantages of having grown up in the UC Davis system. Claude learned by doing. He made a bunch of different wines each year at home, then transferred that experiential knowledge to his work in the small-lot winery at Wente. His focus was pure California, and it wasn't until I started bringing around the wines from other parts of the world that he had any real way of calibrating his winemaking to that done by the larger world.

In the afternoons during harvest, I could always find Claude in the sensory lab, tasting through small vials of wine that had been drawn from fermenters or had just been pressed off. He'd invite me to taste with him, and he'd take a tube holding 50ml of wine, crank off the blue plastic lid and pour half off into each of our glasses. Claude is the guy who knows how to do everything...to fix the things that are broken or to jerry-rig up a way that ends up getting the job done. The earliest pictures of Claude that I've seen show a clean-cut guy with short hair and a well-maintained mustache. Claude is not a tall man but is solid-enough to seem rooted to the ground. The Claude I see every now and again looks like he's been lost in a forest for the last decade. His hair (which he grew out to donate to *Locks of Love*, but which was ultimately rejected because it resembled threshed straw) just continues to grow. The mustache spawned a beard that heads earthward a few more inches each time we cross paths.

He and I would taste and spit and talk about what we saw in the wine, where it might go, how good it could be. This exercise was hugely valuable to me because it got 30 wines a day (that I did not make) in front of me to compare to the wines we were making; to interpret a winemaking philosophy from the brief conversation about each wine and the structure and organoleptics that he was after so that I could put this into my internal reckoning machine and use what suited me and my wines. These tastings would last for an hour or more, would be interrupted by his crew coming in to ask questions about pressing and barreling down, all the things that the *patron* held in his head and dealt out as needed to his men.

Claude became a sounding board and a friend over the years we worked together. Whenever I had a blend that I thought might be the one, I'd ask him if he'd taste it along with me. The answer was always *does a cat have an ass?* This question was the answer to every other question in which the affirmative was obvious, and which rendered the original question superfluous. Like great phrases that stick with you forever after, and that you end up using over and over, celebrating every time you share it with someone else in your life who then adopts its wonderfully off-kilter quality, the cat and its hind-end came to represent a philosophy of life and sharing and friendship that I was proud to be offered.

It seems to me looking back on my time with Claude that it would be a mistake to look at him and winemaking as a man and a problem to be solved (which is the way that I saw him when the forklift wouldn't run or the liner in the press got loose). Fine winemaking became transformational for him. The more he made, and the better the wines got, the more he saw the way they could reshape relationships and link people together. Claude was a brother in the art. It is easy to see him in an abbey in Burgundy a thousand years ago, self-sufficient, and questing; blowing air into his rough-hewn, purple-stained hands as the sun just appeared to moderate the chill. He'd be devoted to his calling for the glory of God then, but the modern Claude always seemed a bit aslant from the university-educated winemakers that haunted Wente's winery every vintage. Early on in our relationship, he seemed to be overly deferential at times to the more traditionally experienced colleagues, at the same time quietly confident in his own wine skin.

I'd occupy part of the sensory lab as other winemakers would be tasting through a bunch of wines and talking among themselves about the utility of micro-oxygenation, adding tannin at the press pan or all the other transactional ways one thinks about when making wine. There are as many winemaking philosophies as there are winemakers. Ultimately, this variety is great for the consumer, as it adds so much choice. Some winemakers enthusiastically embrace technology in the winery and in the vineyard; they cleave to the algorithms and believe that with enough data the answer can be eventually ground out. Others, like myself, believe in a more aesthetic approach to making wine, one open to the liminal moments of transformation that exist for each well-borne, well-loved wine.

I don't think Claude would subscribe, necessarily, to my winemaking philosophy, one of heightened magic, but I think he was more closely aligned with this emotional approach than he was to the data-driven model. Claude retired in 2017, and he seemed happy and relatively unscarred at the party for his going off. I will always look back at that time with him as the beginning of the halcyon days of my winemaking journey. Claude was an anchor to the craft of winemaking; he was my mentor in terms of the practicality of winemaking, the nuts and bolts of how one makes wine and my mentee in terms of what the larger world deems to be great wine. Most importantly, he IS a wonderful soul who has that Italianate-bred connection to and veneration of the concept of *famiglia*.

It is my obligation to incubate great winemakers. I believe that the way we approach winemaking (with an emphasis on truth, on non-interventionist winemaking, on understanding a site and doing only the minimum to keep that expression of site moving, unencumbered, forward) is the proper role of the winemaker. The truest and most glorious relationship that I have during harvest is with fruit. People are complicated; grapes are less burdened. Harvest offers cover. I can get away with less-than-professional responses to email, not return phone calls and the like. Harvest is like a national holiday in this sense. The clocks reset to our own special time zone and the more pedestrian concerns of the day get lost in the tumble of Cabernet tonnage. I love the fact that the grapes own the calendar from August to November. The grapes brook no argument, suffer no excuse; they quietly, and by their mere presence, persistently demand. They are nature's perfect narcissists. There is something of the heroic in doing something, no matter how exhausting it might be, because one must. Harvest is the only part of the business that does not resonate with an emotional timbre. Doing is all. ○

∞

L to R: Aidan Mirassou & Steven Kent Mirassou

∞

∞

June Mirassou

∞

CHAPTER 17

OUT FROM LAND

We finally did move to Livermore a little less than a year after June's initial diagnosis, a year of surgery to remove tumors, chemotherapy courses with pills toxic enough that only she could touch them, the-holding-of-the-breath as we waited for the doctor to show us on her MRI that June wasn't any worse off than six weeks before, and the coming to grips, ultimately, with a too-assured future. Much of 2015 was spent driving together from Livermore to San Francisco to UCSF for drug infusions and clinical trials. We were able to rent the house that rested in the middle of 100 heartbreakingly beautiful acres at the Ghielmetti Vineyard. The vineyard was about 4 miles east of the winery, so I would go to work each day, making wine, writing about wine and trying to sell wine, but I would not venture out to the market as I did before.

Each evening, after I returned from the winery, June and I would pour ourselves a glass of wine in a plastic cup, round up our dogs, and take a walk in the vineyard. Our two dogs were city dogs, though, one of them—Tess (short for Tesla Road—the location of the winery)—was a rescue dog from the streets of Los Barriles in Baja California Sur, but they took to the vineyards as if born and bred. The minute the front door was open, they would tear out of the house running and barking to a metal storage container that the last tenant had left filled with all manner of crap, and paw and snuff around the edges. They had chased a rabbit from under a pine tree on one walk, months earlier, and it jetted under the container. And because of their enormously small brains, they figured that the same rabbit would be making that same carefree journey each and every day the front door was open. That rabbit was not caught by those dogs on that day. Those dogs, in fact, never (in the nearly three years I lived there) caught a rabbit. Capi (short for Capitola) did catch a ground squirrel one day and left it bloody and undone on my bedroom floor, and Tess was 0 for 3 in dustups with the local skunks, but rabbits—not a one. After a minute or so of frantic searching, the dogs would lead June and me down the hill that separated the top of the vineyard from the bottom and which afforded us the most amazing views of western Livermore from our bedroom and back patio.

Most evenings, we'd walk with our wine, hand-in-hand, to a section of the site that we dubbed *The Spot*. A beautiful oak, 40 feet tall, had been planted and surrounded many years later by a clone of Cabernet Sauvignon leaving about 1000 square feet of open space in which we set up chairs and a table. We'd stop and sit, talk and drink, and we could track the dogs from the rattling of trellis wire and the scrunch of leaf as they chased animals. Previous owners had hung a swing in the tree that consisted of a piece of two-by-four, two feet long, secured to the tree with a nylon rope tied through a hole that had been drilled into the wood.

The swing was about three feet off the grassy ground, so you had to put some effort into actually getting aboard the thing. The swing became a shibboleth of sorts, as each new visitor was allowed to pass from the real world into this magical remove only by climbing aboard and letting gravity and a kindly push propel her forward. June and I would hold hands and descend into our chairs for comfort and talk about things small and large, horrible and kind. Eventually, the dogs would return, tongues hanging red and dripping, and we would disengage from that wondrous space and walk slowly back up to the house. In the summer, The Spot was lit late into the evening by the rose glow of the descending sun. The heat of the day would dissipate as slowly as the lover reluctantly leaving his beloved. The sky to the north would transform from azure blue to some shade much deeper, enveloping us in a starry blanket when the sun slid fully into the sea.

We spent a lot of time with our four children, getting together in the backyard that June and I had enclosed in wire fencing so that our dogs could not run off indiscriminately. We'd all sit together and share stories and drink wine together and, in the stillnesses, try to push back inevitability. In March our first grandchild was born, and June was able to spend time with her through Autumn's first year. All living is a series of rhythms. Most of these become ingrained while some (like hair falling out due to chemotherapy, fittings for wigs to hide this, weight loss from the lack of appetite until the drugs bloat you as they kill you from the inside thoughtlessly) disrupt completely whatever sense of comfort and normalcy you usually get from repetition.

    June never fell into the hole of cynicism and hopelessness that I ventured into on occasion. She aggressively sought information about how to fight her disease, vowed to be an *Outlier* (one who ultimately beats GBM) and fashioned an alter ego called the *Mother Fuckin' Brain Cancer Fightin' Ninja Warrior* that she manifested each time a chemotherapy course started. I tried to use our evening walks to quell the mostly unspoken disbelief and sadness that were my constant companions in that time and that blasted me to pieces; June had the ability to just *be*.

One hundred acres is a large space, and when you put 64 acres of vineyard upon its sloping contours, one finds a great many beautiful niches and seldom-seen corners that fire the imagination. On the other side of the most prominent of the arroyos that bisected Ghielmetti Vineyard at the bottom of the site was an olive grove containing 111 trees. Through the grove headed south, you could descend into the arroyo, whose sandy, austere walls were tall enough to cut off the sky. The path that fronted the arroyo ran through the domains of turkey, coyote, and fox. Young kids then—adults now with their own children—had built small forts from wine pallets on the earthen bights that were left after the arroyos dried. Oak and Bay and Eucalyptus shaded the sun and threw out earthy scents of their own. Everywhere you turned on this magnificent vineyard there was another small place that could be fashioned into something extraordinary... a small *fromagerie* here, a little citrus orchard there (you'd never run out of limes and lemons for your gins and tonic!).

One evening June and I were out on the patio watching the sun leak to the west and the glow of headlights descending into the valley from Pidgeon Pass come up in the growing darkness when I told her that I had found a name for this place:

EXULTATION is the going
of an inland soul to sea,—
Past the houses, past the headlands,
Into deep eternity!

Bred as we, among the mountains,
Can the sailor understand
The divine intoxication
Of the first league out from land?

Out From Land was the perfect name. I had studied Dickinson in college and was always struck by this short poem. Certainly, long past the time to do justice to a literary exegesis of the lines at this point in my life, instead, *Out From Land* became a symbol of new beginnings, trials, heading—headlong—into each day with the responsibility to suck the marrow out of it—to live life to its fullest. Often as the sun began to set and pulled the darkness in by degrees from the east, we'd sit on the patio outside, a glass of wine in hand, and watch it descend behind the many ranges on the western sea. At just the right time of night, just before you could pick out the string of headlights making their way toward us from over the hill, you could see each separate range, each wave of mountain as it crested finally on the unseen, watery side. The boat carrying the speaker in the poem, out past the headlands, and into a new day could have been us.

We'd walk the vineyard, our daily circuit measuring about a mile, hand in hand — singularly, the most important moments of my life. Our dogs would run crazy ribbons about our legs as the distance took longer and longer. At the end, when she was eaten through with cancer and unable to walk, I pushed my love in a wheelchair, and the heat of the summer day burned into us the glory that was this fertile, life-giving place. In August 2016, a day before my birthday, my amazing, powerful, gentle wife, immobile on a hospital bed positioned in front of the sliding glass door with a view of the vines she loved, stopped breathing and died.

My wife died at the beginning of the 2016 harvest, and she'll never taste the wines that came from this crop. This particular vintage, this season in time, will always be perfect—a celebration of person and moment, deconstructed (and unperfected) at first by disease but brought whole again by the inevitable grace of nature and Time. Wine bottles time. And though we know of the temporary nature of both, it does a worthwhile job of putting person and moment together. Aidan made a blend of Cabernet and Merlot called *Ninja Angel*, named for artwork that a family friend painted to memorialize June, and it is wonderfully appropriate that his first wine should be the one to mark her last vintage. o

CHAPTER
18

EACH DUSTY SUNSET OBSERVED

The road narrows to one lane as we cross the bridges over the summer-dry creeks. Old scrub oaks with their silvered leaves shade the edges of asphalt and cast intermittent shadows over the road. The trees are old, lichened. They grow gnarled and knobby of their own accord, not hewing to any vision but that of sun and wind and water. It is beautiful out, in the low 80s, as Aidan, Beth, and I drive to a hill-bound vineyard to the west of the town of Gilroy. The windows of the truck are open to the warm wind, and over the music I hear the metallic mewling of a broken A-frame sign moving to the turns in the truck bed.

Up in the hills, through the branches of the ancient oaks, narrow tracks are cut into the hillsides ascending; they are lost behind a riotously leafy screen as we proceed down the curvy road. Brush under the trees waves gently in our indolent passing and is the color of Hemingway's African hills. The seeds of these grasses cavort lightly in the breeze and settle down, like soft rain on the side of the road. The farther we head down the lane, the further back into old California we are. The time shift is complete when we pull up to the gate at Bates Ranch.

Up against the dead-end road is a block of vines marching up the hill; its canes hanging over the trellis wires like the hair of the Woodstock generation, shading green bunches of grapes and throwing a penumbra in a ragged circle around the base of each plant. This was the way they did it decades before in most vineyards of California, before the quest for ultra-ripeness became an end unto itself. "California sprawl" was the way my grandfather trellised vineyards, and for certain varieties in specific growing areas the speckled sunlight afforded by the shaded fruit zone helps deepen the color of those grapes and lessen the degree of sunburn (and off flavors that those damaged grapes contribute to the wine) to which the fruit was subject.

Along the road is a block of Cabernet Sauvignon and another of Cabernet Franc. Planted in the 1970s, the vineyard celebrates its age in the thickness of trunk and obvious balance of canopy. The six acres of Cabernet and four of Franc are collectively called the *Stagecoach* block. When the resort atop Mt. Madonna, which throws its humped shadow against those vines east of it in the late afternoon, was operational in the 1870s, visitors leaving San Francisco and San Jose to vacation in the country would take a stagecoach 80 miles south and stop just there across from where the vineyard is planted. If you love vineyards, which I do, then you dig dirt. On this site, there are three distinct soils that demark the variability of this gorgeous expanse and that will affect the foundations of the wines we will make from them. From coarse loam and red clay along the stagecoach trail to powdery fine sand that coats your boots up at the *Ridge Trail* block, to hardened sedimentary material at the *Luce Block*, each holds water differently and has a different aspect to the sun, and each, in the end, will ripen fruit at a different pace.

Señor Chavez, Nick, pulls up in his Gator to show us around. Chavez is the vineyard foreman and has worked the site for 25 years. He lives in the foreman's house across the road from the vines. As we walk along the rows and talk about the future picking dates and sugar levels at harvest, Nick's squat and powerful dog follows us, then jogs a little ahead to lead. We stop from time to time to look more closely at the vines, and the dog scrambles to rest in pools of shade beneath the vines. At Bates Ranch the three separate blocks of Cabernet Sauvignon run up the mountain from 200 to 1100 feet above the sea. From the main house, each block hangs on the mountain like windblown leaves stuck fast to a wetted window. Señor Chavez takes us through metal gates along wheel-rutted paths that have seen the hooves of a sesquicentennial of cattle, up and up, soils changing from fine sand to rocky loam the further up you go. Each block is more beautiful, more historical, more epic than the previous. Chavez is proud of what his team has wrought in these hills, and he should be. The scales are always tipped to the owners, while the ones who do the physical labor are given the short shrift. Señor Chavez worked his way from farmhand to foreman, from itinerant laborer to the one who has the best views of this part of Bates.

What separates Bates Ranch from the other vineyards we work with is the discontiguousness of the site. Ghielmetti Vineyard, nearly as elevated as Bates at its highest, is a rectangular vineyard on one plane. And though multiple arroyos running through that vineyard affect the direction of planting and the timing of ripening, it is still one continuous piece of earth. Bates is only about 16 acres of vines carved out of 1000 acres of Santa Cruz Mountains wilderness. If the journey up this mountain were like Odysseus' voyage, each vineyard block would be a place of refuge rescued, as from Charybdis and Scylla, from the prehistoric and antagonistic and unbreakable rock of this range. Looking down from the last rows pinned to the summit of the hill, the view is as breathtaking as it is vertiginous. Each block ripens at its own pace, subject to the eminence of a warming sun whose angle of influence is one of the major determinants of that very ripeness. And each block was planted at different times as well. The blocks along Redwood Retreat Road were the first to be planted in 1977, later sections were planted in 2008 and most recently in 2014. As the renown of the site grew, more acreage was planted to provide more fruit to a growing list of enthusiastic winemakers wanting to capture the intensity and profound structure the best vineyards in this area showcase.

We finish our tour and follow Señor Chavez to the main house where a group of winemakers is tasting through wines they had made from the various blocks of the site. These kinds of tastings are always fabulous because they provide a real context for a vineyard of which newcomers like us have no practical knowledge. The most intimate understanding of a vineyard available to man is through a well-made wine. No amount of discussion of the whys and whens of winemaking will ever elicit as much truth as the first sip, the first nosing. That is not to say that the advice of someone who knows what she is doing, who has experience with the site (in good and bad times) isn't an invaluable guide on our journey to make wine that authentically reflects the vineyard.

I know some of the people at the tasting. Prudy Foxx, the viticultural consultant who I worked with on our previous Pinot Noir project; Laura Ness, a long-time friend and wine writer based in Los Gatos. I know of Ian Brand's wines, but had never met him, nor Jeff Emery the owner/winemaker of Santa Cruz Mountain Vineyards, an old brand started by Emery's original boss, and a winemaking legend in these parts, Ken Burnap. There are a couple of other small winemakers here; they had beaten me to this amazing site. Our host is Charlie Bates, youngest son of the original planter of the vineyard, John Bates. Charlie is thoroughly of the cowboy culture, lean and ascetic, traditional hat and belt buckle and a love of large animals that I reserve for wine. Charlie is a great storyteller, and he spins up the past like a DJ making the grand and greatest old hits sound new. Though he is not much older, and grew up like I did when there was more land than houses, Charlie Bates' growing up, separated by only a few miles from mine, was more fundamentally connected to the land upon which the vines and the magnificent family house would reside.

Charlie recounts riding horses out over the original 700 acres of property, hunting small animals, spending summer days that never seemed to end in the warm embrace of youthful confidence and inexhaustible promise. I had land to explore when I was a kid, but I was a visitor on those acres of vine and apricot, taking into me the mysteries of nature and the land only when I could visit my father. Charlie grew up here, spent the wonderfully interminable hours of each season here, took it deeply into his bones, this magnificent world of his. When you are young, there is nothing passive about your existence. Each earthly contour is explored, each hard-won path hacked through thicket, each dusty sunset observed as you finally head home is achieved with the magnificent selfishness of a child.

I mention to Ian Brand, a winemaker making fine Cabernet Franc from these vines, that the wines spread out on the table have a commonality to them, though they differ in age by more than 40 years and were made by different people. The wines taste like the vineyard looks, I said. He said you'd be a schmuck to make them any other way. There are a lot of great places to grow grapes in California, but the best places are those that produce fruit that you cannot imagine being from anywhere else. They have a mark upon them that is so singular and so of the site's zeitgeist that they stand dramatically apart from the mass of good wine as to be nearly unbelievable. Wines such as these that I am tasting today are quintessential examples of the European notion of the supremacy of site over all else.

Jeff Emery brings the oldest wines of the day, Cabernet Sauvignons from 1978, 1982, 1985, and 1989. He also brings some barrel samples of wines from 2017 and 2018 for a new project. All the other wines, brought by the other winemakers, are of recent vintages. Every single wine on the table bears a family resemblance to the other in its blackness of fruit, precision of fruit expression, and shape of the wine in the mouth. There is such a sense of place in these wines; the incredible mid-palate focus and austerity of structure translate the barely subdued wildness of the landscape. There is a perfume to the place that is both aromatic and structural. The trees and shrubs that make up the vast part of this ranch and that run up the slope of hill and shape the intrusions of vine seem to exude summer; you can smell the aromatic oils from the myriad bushes waft over the vines and scent the whole hillside...indeed, seem to scent the very wines made from these grapes. Every single wine on the table has a sense of authenticity and a sense of a very specific *home* to it.

Each wine is delicious. No wine is as profound as the 1989 Santa Cruz Mountain Vineyards Cabernet. Tasting this wine on this day is one of the grandest wine experiences of my life. The context of tasting and what it contributes to one's enjoyment of a wine is nearly always undervalued. No matter how delicious a wine might empirically be, if you drink it with people and food that you love, it will be unparalleled. The joy of being here, on my first trip to the vineyard, with my son and with Beth, both so young and enthusiastic with whole careers ahead of them, on this beautiful day at a site that speaks to me of intent and honesty and truth, tasting such miraculously great wine and imagining becoming part of the winemaking family that takes fruit from this criminally unknown site and works to make something fine from it, vaults these wines and this place and these people into a magical sphere. It is as if I stumbled upon a secret garden in its full blooming and was welcomed, without hesitation, into its embrace.

1989 was the year of the earthquake in Northern California. On October 17th, the San Andreas fault broke near the town of Loma Prieta, not far from where we are tasting, and devastated San Francisco. Just as the World Series game between the A's and Giants was being halted, thousands of bottles of Mirassou sparkling wine were bouncing down the asphalt driveway of their new Champagne Cellars in the Los Gatos hills. For the Bates Ranch site, there is a sense of agelessness. This landscape has been pulled out of history, tamed only to the smallest possible degree. It seems to exist outside of time, relatively unaffected by the machinations of man. Consequently, the Loma Prieta earthquake, that so directly affected millions of people in the Bay Area, is merely a tick mark on the geologic timeline here on a hillside in the Santa Cruz Mountains.

The wine that Emery made is, above all else, delicious. The wine is still young (in fact, all the wines—even the 1978—belie their ages), no hint of the oxidative aromas that old wines are heir to…overripe fruit, and a general tiredness. This wine is youthful, vibrant, contained, precise, long in the mouth, compelling in structure, showing a broad range of flavors and complexity but also hiding much still. I taste through all the wines but come back a couple of times to the 1989. It is one of the few wines that I've had that I wish I had made.

Harvest is a short time ahead, and I arrive this time at the vineyard gate on Redwood Retreat Road about noon. It is beautiful out with skies impossibly blue. There is nothing like the mundanity of clear skies after several weeks of smoke and haze building a ceiling of blood red over your fruit to recognize the everyday miracle of cloudless azure. I am alone today and excited to get a chance to explore the estate and try to learn, by walking and tasting, the language of Bates. I'm also here to pick the rows from which our fruit will come. I am a believer in the sanctity and specific quality of a site. Not every piece of ground translates the details and idiosyncrasies of terroir in an understandable and meaningful way. Often these vineyard-specific wines are muddled or lack a voice; Bates Ranch is not one of those places. It, instead, speaks of altitude and unencumbered sunlight, folded up in the ridges of mountain like a flower's petal balled up in a white napkin. I get in my truck and drive the rutted, cow-blocked road up to the top of the ranch, figuring I'd work my way down to the road.

The Luce Block, named for a family member of Charlie Bates, is one of three small sections planted in the sand near an avocado tree, pregnant with (disappointingly) unripe fruit. The top block is planted along the hill while the other two run unhaltingly downward, showing a 90° different face to the sun than the one above. The block is split in the middle, so I walk off eastward to one end of the block tasting berries and trying to absorb the whole thing. It is warm today, but intermittent breezes from the sea wash up against Luce and bathe her in a salubrious exhalation of cool perfume. There is mintiness there, but woodsy and not from the garden; there are notes of citrus too—more orange than yellow—and as I rub my hands through a woodsy, oily bush, they are anointed with the spices of a foreign bazaar. I am afloat in a warm and welcoming sensory bath.

After walking up and down and across the hill here for about an hour, I have a sense of the rows of fruit and how they will fit into our wine program. I don't really know anything yet of the soul of this place so fragrant and epic, but I can tell differences of structure and taste and how each row of fruit might express itself in a finished wine. I put duct tape labeled with *The Lineage Collection* on the endposts to designate our choice of location and move on to the next section.

Ultimately, someone has to make the call. I am armed with too little experience, I realize, as I chuck berries into my mouth, to justify what I am spending for each ton of grapes. Nevertheless, I need to divine what I can, compare the flavors and structure to the millions of grapes I've tasted over my career and fashion a vision, ethereal as it may be, of what these grapes hanging pendulously here, now, are going to taste like as wine three years hence.

The upward cant of Luce brings out the sweat, Ridge Trail braises me in my own juices. This section of vineyard was planted in 2014 from cuttings from the original planting at the *Stagecoach* block. And it was planted straight up and down. Walking down the slope of this block is difficult as broken rock juts out from the ground, and a skree of clay works like ball bearings under my boots. Walking up the next row is like… the slog up

Everest. I do make it to the top, with a stop or five, to slow down my heart. My shirt is soaked through on the back, but I have a better sense of how grapes ripen as the sun cuts the vineyard on the transverse. Señor Chavez tells me when I get down to the Cabernet Franc, later in the day, that the labor contractor sent elderly men, one year, to pick this block. One guy broke his boot in half trying to walk down the hill, and another had to use the trellis wires to pull himself up the hill with his lug box. The fruit came in much later than the winemaker was expecting.

Great vineyards, like Bates, exist on the margins. Nothing about them is easy. The soils in certain parts, like the Ridge Trail, don't hold water so there is uneven ripening. Getting fruit out of the vineyard is a pain in the ass. My trucker, with his 32-foot gooseneck trailer, got stuck trying to get out of the postage stamp of a staging area and had to be assisted out of a ditch with a forklift. The loading took twice as long as we anticipated. None of that matters, because the fruit is more than good, more than potentially great; the vineyard itself is that special and rare nexus of geography, aspiration, and challenge whose fruit has the potential, with each opened bottle, to reconnect the winemaker to that first time he laid eyes upon that dazzling mountainside. For the wine lover, it is that place and time that defines a lifetime's personal search for comity and beauty and for a wine that strums a revelatory sequence of notes that cracks open our hedonistic and thirsty heart.

My trucker charges me $900 to deliver a load from Bates, so I rent a flatbed from the local equipment rental place in Livermore and drive down to the vineyard to get the final four boxes from the site: 3 from Stagecoach, the block along the road, and 1 from Luce 2; two tons. The three from Stagecoach are destined for one 1.5-ton fermenter and the lone .5 ton from Luce 2 going into a picking bin for the alcoholic transformation. I decided, after walking the rows again, on a Monday before the Thursday pick, that the ravishing qualities of flavor and structure in the Luce 2 block were better than what I was tasting from Luce. So, my internal calculus is one of trading the potential complexity of Luce with its old vine planting, tasting green now and needing a week or more to be optimal, with the sublime flavors of Luce 2 (I had already harvested a half-ton a week earlier and was even more excited about the destemmed and crushed fruit than the glorious stuff hanging on the vine) and the real-world exigencies of harvest schedules and the availability of a fork-lift to get fruit onto the truck. The fact that I'm here to pick up a ton-and-a-half of Stagecoach Cabernet, which was planted in the 1970s, undergirds my harvest decision. As I'm walking the rows to cement my sensory impressions, I'm talking to myself about as many of the *whys* and *wherefores* as I can…the inevitable tradeoffs that are the currency of non-estate grown fruit as I formulate the harvest plan, in this case, on the fly. I talk with Prudy Foxx, the estimable vineyard consultant and a friend, about my dilemma and she sketches out a scenario that aligns with mine. We make a plan via text, and eventually (there is no service at the vineyard until you are at the very top overlooking the entrance to the Salinas Valley to your left, scores of miles away. After a few mis- and un-communications, we finally confirm. From the top of the vineyard, looking south, there is a wall of gorgeous evergreen rolling up the mountain on the other side of the valley. Given the intense fires that are raging in Napa and Sonoma counties, it is hard not to imagine the disaster that a wayward ember would cause here.

Being able to work with this fruit is exceedingly exciting because it is so singular. The difference between the wines from this site and Livermore are a 10 on the scale. The difference in fruit from Napa and Livermore is only a 5. As a lover of—and longer for—those wines that carve a definite place and time, the Bates Ranch site is the quintessential example. Unapologetic in its lean-ness (relying upon it as the authentic mark of the site), I am especially concerned about being able to make a wine that expresses my awe of the site and that falls within the family of wines that have been made historically from the vineyard. There are certain AVAs and individual vineyards that are compliant enough (read, that lack compelling specificity) that the winemaker can manipulate the fruit in a way that makes a generic wine reeking of winemaking only, that has no connection to site and authenticity, but which sells in a marketplace that trades in mundane generalities. There should be no way to mark the wines from this site as anything other than *Bates Ranch*. The wines argue for one interpretation only, unriven, bone and sinew from the chaparral and the oak and the cow and the open sky, cleaved to the chucked and vertical road leading to the bluest sky, holding within them an Old World immortality, a sense of completeness, a sense of remove (an inurement to critical voice and marketplace place-holding) that slots them into a category whose parameters are adamantly and exclusively *Bates Ranch*.

I nose my 24-foot truck into Nick Chavez's driveway and turn around so that the four bins can be loaded onto it and I can drive away with a minimum of K-turns. The one bin of Luce 2 takes the longest because the pickers grab fruit in an orderly fashion from one block at a time (and I'm the only one getting grapes from the uppermost reaches of the site today); the three half-ton plastic bins come off with little difficulty, my rows coming to me and other grapes going to other winemakers.

My truck is loaded, and I tie down the bins and head back to the winery. I stop at the corner burrito stand just outside the vineyard, pay the owner through Venmo, and head back to Livermore through the pall of smoke blowing down from the Napa fires, steering with my knees as I eat my *carnitas* burrito, heading north.

A week later, the wines are approaching dryness, and they are showing an expressiveness of flavor and texture that pay back every hope we had for the grapes. Black fruit, a dusty underlayment of dried herbs, significant tannin; the wines are reflecting our perception of the fundamental nature of the site as any great wine must. Rusticity, and Old World seriousness pervade the fermentation bins now, and we have great hopes that these wines will mirror, when they are finished, what we are tasting now. ○

∞

Nancy Castro, Xongo (the German Shepherd)

∞

CHAPTER
19

SHE IS BEAUTIFUL AND HAS BROWN SKIN

I'm headed over to a release party at a buddy's winery a mile or so around the corner from my own. He is about to celebrate the birth of the newest vintage of his finest wine, and I want to support him as well as to taste what the competition is up to. Many of his club members also belong to mine so entry into the tasting room is full of hellos and what-do-you-thinks of this new offering. A woman comes up to me to check me in and give me a glass. She is beautiful and has brown skin that is deeper than just sunstruck. She asks me my name and makes a mark on her page. She strikes me as being extremely self-possessed; I don't know her, but I can tell that she has had jobs before like the one she's doing now. She tells me about the wines that are being poured and where they are and wishes me a good evening.

I proceed to the first table and put my nose into my glass, telling what I can from what I'm smelling. I'm looking over the rim of my glass, too, to see where that woman went. She comes up to me at the next station and asks if I know a good friend of hers. She shows me a picture of her, but I don't immediately connect her to the person I met a couple of times ten years earlier at an annual tasting I poured at in Monterey. We talk more, about her job, where she came from, the little things that, perhaps, thread together a more durable cloth. Nancy Castro, I learn, is the new Tasting

Room Manager, she has been working in tasting rooms in Sonoma and Napa for a couple of years and has just made her way down to Livermore, a place she did not know, to take up this new opportunity. I taste through a couple more wines, find Nancy one more time to say good-bye, and email her the next day to see if she'd like to have dinner with me.

The following week we're at *Sabio on Main*, one of our great local restaurants. My friend, Francis Hogan, partner and chef, comes by the table and tells us we have to have an order of barnacles that came off the boat that day. *Barnacles? Bottom of the boat, barnacles?* Francis says yes, bottom of the boat, and they are delicious. We can't pass that up. Another friend comes by to give me an old bottle of Mt. Eden Cabernet. Jeremy Troupe-Masi, host of a local podcast and a true believer in the potential of the Livermore Valley, worked a harvest for Jeffrey Patterson, another friend, and the owner/winemaker of a venerable brand in the Santa Cruz Mountains, and he knew how much I love Jeffrey's wines. It is a very generous gift, and a wonderful wine. Rich and balanced and elegant, replete with the rangy textures I associate with those mountain sites, black fruit, and a confident if not brash structure, Nancy and I drink the bottle down over the course of a couple of hours. The barnacles come, and they look like mussels with a leathery penis hanging out of the shell. Francis tells me to take the skin off the foot and to bite off the pale meat that remains. This is a lovely conversation for a second date, but I do what I'm told. My first effort leads to an eruption of barnacle splooge that stains the white oxford shirt I'm wearing, but the meat itself is amazing. It's like the most evanescent combination of the briny tang of oyster and the sweet, dense flesh of Dungeness crab. The bowl doesn't last long.

Our conversation tonight takes us all over our personal geographies. What it was like working in Sonoma, what movies do you like, have you ever been to Europe; the casual questions whose answers, nevertheless, add color and texture to this person's inked outlines. Time passes unhurriedly and deliciously, and the present (there in the restaurant, with the clatter of the dishes and the calls from the kitchen and the martini shaker keeping time to our conversation) gently expands to contain my sense of possibilities. It seems like no time at all that we are the last ones there, and I'm a bit embarrassed, knowing how hard the team at the restaurant works, to have kept them. I drive Nancy home and know I want to see her again.

Nancy is incredibly driven and has as her guiding business principle the responsibility to create from the company something self-sustaining and great. She feels she has something to prove, and I feel as if I have found a comrade-in-arms, one devoted to chucking the status quo aside and to affirming our love for what we do and what we can accomplish together. In many ways, she is the strength that I find myself lacking. She can make the hard decisions, the necessary ones. I'm a bit of a pussy in some ways. Perhaps I want, too much, to be liked; perhaps, I have given to inertia what is inertia's or have mischaracterized festering challenges as mundane when their solutions are, instead, characteristic of true leadership. Because there are no vacuums in business, I have sometimes abdicated an authority to others in my organization that ends up getting exercised by them because daily business issues need to be dealt with. That they are often dealt with in a desultory way that lacks subtlety or vision serves to compound my original failure. My inability, then, to lead and to persuade comes from years of leaving an operational and philosophical void, filled by whatever day-to-day exigency that exists out of my vision. Nancy is the one who shows me how this rots the floorboards out from underneath a company's true mission.

I bring Nancy out to *Out From Land* after our first night together at her place. This is a sacred place; it is a test in its own way also. If you are the right kind of person, there is no way to be inured to its influence, to its magic. We are on the verge of spring, and it is warm outside. She and I head out with my two dogs in the late morning to take the loop that defined so much time and progress with my late wife, June. Capi and Tess take off running after rabbits just outside the front door, and we head down the hill that is one leg of a capital *U*; the other leg marks the entry into the lower half of the vineyard. Nancy is the only woman other than my late wife to walk this path with me, and the significance shakes the ground. She doesn't know yet about that ruinous year of incurable disease, made slightly less horrible by the daily walks along this same path. This path, no matter how magical, no matter the fact that I glean that Nancy senses the enchantment that infuses this place, the perfume of possibility, the ode to potentiality in every screech of bird and rub of trellis wire, the sorrow and loss of that year is here as well, cutting an uneraseable rut deep into this fecund ground.

About half-way in, we pass *The Spot* (perhaps the most sacred of places), and Nancy stops to take in the small and divine area in which June and our kids spent many lovely days being with each other. She clenches my hand again and her empathy begins to fill up that black hole into which I had sunk and begins to bring me back up level. The dogs lead us (their route far more extemporaneous than ours) back up the hill to the house. The days are short still, and the sun is setting as I pour us a glass of wine. The western view is as it ever was, full of import, full of possibility, epic, and perhaps, now, full, too, with another person to face the time ahead.

A couple of months after our first walk around *Out From Land*, Nancy moves in with me. She is the perfect companion. She understands implicitly the draw of this place, relishes it, and perhaps it fills in holes for her as well. None of my visions for the place meet unenthusiastic ears. They redound, in fact, to her even greater plans. Every day we walk our circle, sometimes veering off together to parts yet undiscovered. Our first foray to the lower arroyo, through the Olive Grove, leads eventually to our harvest and subsequent pressing of the gorgeous olives that have littered the ground for years before her arrival. We cook with that oil still.

We are in the spring of possibility. Each niche of *Out From Land* can be used to bring pleasure to people, to make their lives richer. We can do weddings here, amazing wine dinners under the stars in the vines there. Nancy arranges for the local wedding rental company to do a photo shoot in the barn east of the house, and it goes off beautifully. Our enthusiasm for the marriage of this gorgeous place to the intimate desires of potential visitors grows. Our plans bloom, many of them hatched as we walk our circuit, wine in hand and dog at feet. This site becomes our place of magic. Each time we'd turn on to Reuss Road at the bottom of the hill, coming back from the winery or from shopping, the vineyard would roll down to meet us, to draw us into its verdant bosom. Each contour, especially in the waning light of late afternoon, painted by the last orange of the sun before it was lost to the sea, held mystery and comfort both. *Out From Land* attached itself to our hearts and our imaginations with durable roots as it became one of the main characters in the story we were writing on a daily basis. Looming in front of us, unknown as of yet, were the last echoes of the discordant chords of a financial disaster begun in 2008, a bank's prosaic shedding of assets, and the final tearing out of those durable roots and the falling-in-upon-itself dream of our vineyard life.

.... ii ....

In the beginning, the Ghielmetti family invested in The Steven Kent Winery as a way of helping to integrate the production side of the business (wine-making and vineyard) with the marketing and sales side. They were always supportive and generous partners, and we agreed from the beginning that they were not going to be invested for the long-term. The money they invested in the company took out my original partners and gave me the opportunity to trade debt for equity with the profits from the company. We had fluctuating levels of success over the nine years that we were partners, and just as we seemed ready to really take off, the economic crisis in 2008–09 knocked out the supports of the high-end wine business and stopped our upward trajectory cold.

When my partners' investment period came to a close, they generously offered me the first opportunity to purchase the vineyard. I spent harrowing months trying to put a deal together to buy it, but in the end, it was acquired by another wine family in the valley that had made a fortune selling tours to Hawaii. The loss of the site was unfortunate enough, as we were losing control of the finest fruit source in our appellation. But what made the fruitless months of deal-making so personally devastating was the knowledge that I could not succeed in being the steward of a site that had so quickly become an emotional center for my family, the place where my kids' mother's ashes were spread, and where we all saw generations of the family abiding.

Nancy and I lived at *Out From Land* and supped of its natural beauty and its symbolic possibility for just over a year before we had to leave. We spent the last two months looking for another place to live and packing up boxes. We put clothes in suitcases, moved our table from The Spot, took all but two of the Adirondack chairs off the patio and stripped away all of the physical accoutrements that signified our love for this gorgeous place. We found a house to rent in the neighboring town and began transferring our stuff. There was an overlap of a few weeks so we spent as many evenings as we could up at the house on the hill looking westward. Our last night there, I raced the setting sun, down Tesla Road so I could raise one final glass of wine with Nancy to our foreshortened adventure. The sun was fully round and hung at the dusty horizon like the ripest orange. The day, which had been very windy, cooled us as we loaded fruit into the press earlier, but now it had stopped completely, and the night-time sounds that so often accompanied the music of living up here, were, like the passing trains in the valley, shortened and shifted. The western mountains bit into the sun and devoured it quickly. The vines below us, still and pregnant with fruit, harbingers of dreams of their own, rolled down the hill in green bands, then dissolved quickly in the blackening day. The structures that commanded our view from the western patio disappeared too, only the twinkling house lights on the margin making the hidden buildings manifest. With the darkening came the cold. I grabbed sweaters for the two of us, and we held hands as we sat on our Adirondacks drinking down the bottle to the bottom. We turned out the porch light and our backyard—the little fairy greenhouse, the lime tree I planted to commemorate a year of June's survival of GBM, the hammock that held me and all three of my dogs during the slow and balmy days of summer—dissolved away too. We went to bed, and the next night our evenings began to unfurl in the cracked-concrete backyard of a tract house in Pleasanton.

Nancy shares my bitterness, scabbing over now, at not being able to make a go of *Out From Land*, but she shares my essential optimism too. Each of us wants to find that place that becomes the centerpiece of generations of family, home base for kids and dogs. The loss of the vineyard has not dampened that desire, certainly. The whole experience at *Out From Land*, the death of my first wife and my bereavement and my being brought back up to the earth again by another woman I love dearly, has underscored the fact that we never know how much time we have to fashion from the world that which brings us the greatest happiness. Time is no friend and must be vanquished, if only temporarily, by activity and industry. It is in our nature to reshape current reality into a more wholesome and magical presence. It is what we both do and having Nancy there (she who guides and goads, nurtures and inspires) makes the journey so much less lonely and more profoundly resonant. ○

CHAPTER
20

THE CLEAVING OF SOUL TO SOUL

Nancy and I spend the 4th of July with close friends, Matt and Rebecca Toomey and their kids, and Beth and her husband, Patrick Refsnider, at the winery. It's lovely day full of laughter and wine and food. And just as the sun sets, we make our way to the top of Sachau Vineyard, one of our preeminent sites for world-class Cabernet. The heat of the day has mellowed, a soft breeze comes in from the Bay, fireworks wink like luminescent bugs down below in town.

We watch from the bed of a pickup—drinking great wine—the kids race the dogs up and down the hill. The full and beautiful moon then rises like a spotlight behind the hills to the east and we all lose our breaths and are compelled by the same otherworldly synchrony of being there at that place at that time with that group of people knowing the specialness of the moment and knowing that it was likely those young kids would recount (like my grown kids do the magic of Lake Powell or the simple shoulder-loosening wonder of the beach at Capitola) forever, that glorious night when the troubles of men were washed away by the soft, ministering hands of nature, and love was abundant.

If the arc of the harvest is like a long and fraught journey away from the safety of home, full of unexpected stops and turns-upon-themselves,

July 4th marks that moment when you can see, once again, the recognizable contours of a familiar coast. It is before veraison (when grapes start to color up and grow fatter and soften) but the berries are recognizable for what they will become in little over a month. We'll be pulling Sauvignon Blanc into the winery by mid-August and the earliest reds, like Sangiovese, will follow as September turns. To see the vineyard as it is now, settling into the earth under the lowering light, smudges of leaves and shoots, is to feel both protective and protected. These plants have weathered millennia and have switched off and on their genetic toggles to arrive at a point at which their needs intersect so beautifully and ineluctably with mine. I harvest their offspring to change the lives of my species, acknowledging late in the night, typing these words on my patio under the influence of gin and Miles Davis, that my actions also guarantee the perpetuation of their species in every celebration my wines engender. There's the protection part of the equation. This beautiful canopy of green that just seems to pop up each year almost overnight to cover the rocky and alluvial earth and to sup from that earth too, to transform the old, flaky trunks into a verdant bower and to give me the fuel each year to reimagine and reinvigorate my artistic life is the stuff of Proserpine. That's the protective side.

There are discreet moments in which the state of the vineyard mimics my mental state. This night under that gorgeous moon with those loved ones and the warm and rich and comforting breeze settling us so deeply and comfortably into that site is one of them. The magical night under the July sky lasted only a few hours in normal time. It's effect on me, however, is of the timeless sort. There is a sense of completeness to the experience, as if we are all meant to be there, the moon full and bright; the colored gunpowder popping down in the valley below; the wine flowing succulent and endless, bringing us closer to each other, closer to the burgeoning, impending, vineyard atop the Sachau hill, among whose expansive and beneficent and enveloping verdancy, we all came together.

.... ji ....

Wine is food. Food is necessary to live. Wine is, therefore, necessary to live, for those so inclined. You don't always need filet mignon, however. We are in existentially weird times now in the wine business. A virus rages outside our doors like the hordes from a far-off land. We are all stuck inside our houses hoping the enemy tires or is driven off by superior forces or superior firepower. Homes are bubbles, businesses are closed, the confederacy of dunces in the capitol can't get out of their own way. People continue to drink, however. And bless them. Gigantic disruptions like Covid and the 2008 housing crash before it have profound effects on the wine market. High-end wine is a luxury. When the market is in the toilet or when jobs are at risk, high-end wine lovers move down a few notches in price point to drink as well as they can. This understandable phenomenon impacts brands like mine. We do not make any real substantial quantity of wine below the $50 per bottle mark, but necessity spurs us to invent. The wine world today offers consumers an amazing array of opportunities to drink well at reasonable prices. We will, on occasion, play the bulk wine market to find really nice wines at terrific prices, blend them with great barrels from our cellar and create proprietary wines that work on the palate and the pocketbook. Being able to navigate this market allows us to connect to more wine lovers as well.

I have a profound desire to bring a delicious bottle of wine to every table in America. The youth of the wine culture in America, especially outside major markets, sees wine as something *other*. We do not see, yet, the natural affinity of wine to food, the millennia-old relationship of the food you rip from the ground and bring down with arrow to that you create through fermentation. When the United States is at a point that we celebrate the simple act of everyday eating (food and wine playing the roles of yin and yang and real time spent re-connecting with our family), we will have become, then, a mature culture with its priorities in order. Under the best of circumstances, the wine consumed in this dynamic will not need to be expensive. It will only need to be heartfelt and delicious. Like the heirloom tomatoes that are lovingly coaxed into a sauce for pasta, or the flour that is turned into that sauced dish, wine can be an ingredient as utilitarian as that tomato and that ground wheat in its role of marrying with and showcasing the deep connections to honestly made food. At the same time, wine—unlike the tomato and the wheat—has the power to deepen and transform the relationship of the people to the food, and the people to each other. This is, perhaps, wine's greatest gift, the cleaving of soul to soul.

Though we are the country that banned the production and consumption of alcohol for twelve years, and neoprohibitionistic forays continue to pop up periodically like a *whack-a-mole* game, especially when those who want to take away think the other 99% are having too much fun, the business of wine may be, finally, too big to fail. In the last ten years, Supreme Court decisions regarding interstate commerce have led to the opening up of the majority of the country to direct shipment of wine. The few straggling states are as anachronistic in this regard as they are with most issues related to making life better and more equitable for its citizens.

It is not government fiat that will bring about the proper relationship of people to wine and food; it is time. It is the passionate work of winemakers, too, working apart, working in their small wine growing areas, and out of their small wineries, that will root the grass upon which, over time, and by dint of habit, the culture of wine and food and living more fully will find cool bedding. The parents enjoy wine on a daily basis, and their kids observe the rituals. They hear, at times, the conversations between mom and dad about how delicious the wine is and how well it goes with the chicken. They grow bold and ask for a taste. The responsible parent puts a few drops on their tongues. A small few will immediately have the bells and lights go off, and they will be bound deep into a profound and deep relationship with wine from that point forward. Most of the others will find the foreign flavors and textures too mystifying to grab ahold of at that point. Some will never find the taste for wine, and their meals will reek of dry sands. Many will persist and will eventually break through. And they will come to understand the simple truth that wine adds greatly to a meal, to food, and thus to life. o

CHAPTER 21

AGELESS HEART OF THE WORLD

I get off the plane in Oakland after a week of selling in Southern California, a little tired and mentally beat up. I lug my bags to my truck in the parking lot and head to Livermore. I drive directly to the winery to check in with my team, change out of slacks and wingtips and into jeans and boots and head to the vineyard. I get out of my truck at the top of the hill at Ghielmetti, look down west to the sun moving seaward, and feel my lungs and heart expand. It has only been five days, but there is so much new growth that I could be waking fresh to Brigadoon, to the magical exhalate of a dream.

Fruit hangs in the waning shine of summer, hearty and sweet. The bees are buzzing gently around the bunches, alighting upon broken berries, drinking from them greedily, making honey from the same sugars we make our finest wines. I walk carefully down the rows, a little overwhelmed, reach into the canopy every now and again and pluck a grape. The fruit bursts, and I feel the heat of the early day suspended there in the nearly ripe, viscous juice. The bees are welcome, as they are comparatively rare and doing important (and delicious) work. The Starlings, massing in synchronized clouds above me, are another thing entirely. I hear them in the silence of the afternoon come up through tangled vines to crack berries open and sup upon the juice. At the end of rows, mostly, I see evidence of their marauding in jagged tears like acne scars on the shoulders of bunches, and I go to lengths to limit the damage. In the better vineyards of Pinot Noir, nets are used (pinned together with plastic spoons at the bottom) to clothe the fruit zone; in others, mylar strips tied to one long cane flash in the sun, kites that remind me of Emerson's transparent eyeball resist the facing breezes above the vines, air cannons pop off intermittently, and the recorded notes of raptor erupt from the inside of the rows in the warm and lowering light. Even amidst the cacophony, the birds get their share. I walk a few rows and get my bearings. I am in one small piece of the vineyard, but I can feel the entirety of it, all 64 acres, the gentle immensity of it pushing up against me and enveloping me in the advancing aromas of warm straw and broken grape.

.... ii ....

Certainly not just the fact I'm over 50 now, or a grandfather of two perfect grandchildren, but I'm acutely aware of time passing and all too aware also of the fragility of life—the lives of loved ones and of companies. Success is taking on a different meaning for me now. While it can never, nor ever should be, uncoupled completely from the inevitability of commerce, it is the pure, more fundamental relationships that drive me now. My idea of success for my brands is being circumscribed evermore completely by relationships. My idea of the proper relationship between the winemaker and the vineyard and the craft has changed dramatically. No more am I trying to twist and mold and lengthen and compress. Now, I'm in a symbiotic relationship that is less about control and *making* than it is about revelation. Each element works in concert with the others and the process of sharing energy and desire with all my "partners" can lead to moments of perfection or a state of perfect *now-ness*.

I am happy that I've realized this now so that I can spend the rest of my career unlayering complication in myself and in my wines in order to approach true clarity. Each little stage of this process—a moment of seeing my work for what it is and what it means—is illuminating and makes me more whole. As I now can see, success, for me, is seeing my wine club members and my guests at the winery come away from their experience with a deeper understanding of the utilitarian mystery of wine and of the place that wine holds in a fully-lived life. Without an avid and satisfied receiver of experience, the most magical moments echo insignificantly in the vast nothingness of today's techno-reverb. It is my hope that the deep connections we are trying to forge will invite those who love wine to contemplate a more authentic and intimate experience. When this happens—even if only rarely—that will be a deep and good thing.

The final piece of my evolving vision of success is the caretaking of our land. There are multitudes of sources available to describe how one can farm organically or biodynamically, and they have profoundly affected my thinking about what we are doing now and where we ought to go. The simplest way for me to describe this relationship to the vineyard is embodied in a phrase you hear if you've ever hiked in the White Mountains. "Pack in. Pack out." Simple. Don't leave footprints. Leave things the way you found them if they're right. Make them right if they're not. It is crucial—as with every other relationship—to work to reverberate at the same frequency so that the outputs naturally come from what has been put in. Perfection doesn't mean a lack of flaws, and it comes about by doing just those true things that need to be done. The examined life is spent discovering what those true things are.

I'm a big believer in being open to the everyday wonder that makes living day-to-day full of wonder and potential. For most of us the 8 to 8 is full of just getting by, getting the bills paid, making sure the kids are on the right path, that spouses are connected. All of these things are important and create the mesh which binds over the disconnected parts to give our lives value. But it is the interstitial moments, those cracks in the delicate frame we try so hard to reinforce, that *possibility* peeks through and the magical moments come up a-blossom to the surface.

.... iii ....

There is no more beautiful landscape than a vineyard at its most *full*. There isn't any other greensward with more promise, more mystery, and more hedonistic delight. Where I am now, higher up, I can see the ordered rows laid out as they run away from me, hugging the curve of earth to a single point at the far margin, rays of green leaves greedy for light, laden—heavily pregnant and promiscuously heavy with the stuff of '45 Mouton and '61 Cheval Blanc. Though I get a sense of the continuity of the family business when I'm in the cellar or on the road traveling from one market to another, I feel it most keenly here. I'm not sure why. I grew up away from the vines and became a man in big cities, but it is in the middle of the vines, alone and perceptive, that I feel most consequential.

It's in our vineyard this evening on the eve of the harvest when the last light comes in on the horizontal with all that life surrounding me, that I realize there is no finer place for a summing up of life to this point. This vineyard is the place where my wife's ashes are spread, as it is the place I fell in love with my fiancée. It is the synecdoche for all things of humble and perfect worth just as it is our family's *Ur*-site: in the fading light of the day, if you squint just so, you can see this place shimmer and become the vineyard my thrice-removed grandfather first planted 160 years before in San Jose. In the soft warmth of the rising night, they have become one and the same, planted by the same men (the connective tissue of Time raveling in both directions), finally meeting after more than a sesquicentennial adrift and apart.

There are inevitabilities: empires will fall, and the ones we love deeply will pass. The rows will keep going forever, though, meeting down at the horizon beyond our sight. The fruit will be suspended in the joyful and warm and full yellow of summer always. The winemaker's shadow (his, mine, my child's and grandchild's) will hang out long against the earth in the late hours, and the life around us in this refuge will vibrate and buzz and fill us to breaking. We will kneel in the rows in the warm shade of day's end and be clothed by green leaves, and we will plunge our hands into the fecund soil as we feel the ocean-pushed breezes cool our shoulders, and we will know again that we are ineluctably rooted to the green and ageless heart of the world. o

ACKNOWLEDGMENTS

I did not know if the words you are reading now would ever be written. Bits and pieces of this work have been jotted down over the years more as a beacon of hope when days were their toughest than as something tangible that I would eventually be able to hold in my hand. If not for a sharing of the manuscript with my father one day, who shared it with a friend, Tony Kilgallin, who then shared it with editor, publisher and friend, Paul Chutkow, it would have stayed only a monument to intermittent creativity and desperation. Thank you, Paul, for your constant enthusiasm for Lineage. This book is yours as much as mine. I'd also like to acknowledge the support and love of my fiancée, Nancy Castro, and my amazing kids, April Coffey, Aidan Mirassou, Katherine Mirassou, and Sara Mirassou. All of you inspire me every day to reach for large goals. Finally, I want to acknowledge all of those, like me, who find purpose, fulfillment, and creative joy in the world of wine. There is no finer way to bring richness to the lives of others than through this wonderful act of teaming with Nature to produce a product so profoundly alluring, historically important, and flat-out delicious. Cheers!

INDEX

Alexander, Mike, 102–103
Alta California, Mexico, 14
Arroyo Cellars, 17
Arroyo Mocho river, 71

Barbera, 64, 72–74, 113
bacteria, 51
Bankowski, Kate ("Klondike Kate"), 69
barrels, properties of, 109; 121–126
    French vs. American woods, 122
    toasting, 123
Bates Ranch vineyard, 151–160
Bates, Charlie, 154
Bates, John, 154
Bent Creek Winery, 17
Big White House Winery, 17
Biodynamic wine, 132–133
Boaventura de Caires Winery, 17
Bobba, Claude, 139–142
Boughner, Matt, 7
Bourdeaux varieties, and Livermore Valley, 14, 18, 38
Brand, Ian, 154
Brettanomyces (Brett), 124–125
Burnap, Ken, 154

Cabernet Franc, 59–66, 152, 154, 158–159
Cabernet Sauvignon, 18, 41, 42, 50, 52, 55, 60, 61
    Bates Ranch, 152
    Concannon cuttings, 74
    Home Ranch, 74
    Mirassou 1968, 89
Caddis Winery, 17
California wine and Paris tasting (1976), 19, 34, 61
California, history of winemaking in, 34
Castro, Nancy, 73, 163–169
Cedar Mountain Winery & Port Works, 17
Château Margaux, 74
Château d'Yquem, 74
Champagne (see also sparkling wines), 98–99
Chardonnay, 40, 41, 49, 52, 61, 64, 99, 104, 113, 116
Charles R Vineyards & Winery, 17
Chavez, Nick, 152–153, 158
Cherry Blossoms Sangiovese, 45
Cheval Blanc, 1947, 39, 51
Clarin, Mark, 139
climate, California, 72–73
Coffey, April (Mirassou), 25, 117
Concannon Winery, 17
Concannon Clone 7, 74, 113
Concannon Clone 8, 75
Costanoan people, 14
Cresta Blanca Winery, 15, 18, 74
Crooked Vine Winery, 17
Cuda Ridge Wines, 17

Dante Robere Winery, 17
Darcie Kent Winery, 16, 17
Doerschlag, Mark, 85
Dom Perignon, 99

Eagle Ridge Vineyard, 17
Easley, Chuck, 106
Ehrenberg Cellars, 17
El Sol Winery, 17
Elliston Vineyards, 17
Emery, Jeff, 154, 155

Fazio, Jennifer, 70, 79
Fenestra, 17
Fiorentino, Marcello, 72
Fisher, Janice, 79
food, and wine pairings, 85–90, 171–174
Foxx, Prudy, 154, 159
Fuezy, Ivan, 41–43

Garré Vineyard & Winery, 17
Ghielmetti Estate vineyard, 5–7, 23, 108, 124, 145–149, 167–170, 175–177
Ghielmetti family, 167
Gilroy, California, 151
grapes
    Barbera, 64, 72–74, 113
    Bourdeaux varieties, 14, 18, 38
    Cabernet Franc, 59–66, 152, 154, 158–159
    Cabernet Sauvignon, 18, 41, 42, 50, 52, 55, 60, 61, 74, 89, 128–129, 152
    Chardonnay, 40, 41, 49, 52, 61, 64, 99, 104, 113, 116
    Mourèdre, 34
    Sangiovese, 42, 50, 71, 113, 171
    Sauvignon Blanc, 6, 41, 61, 63, 74, 75, 171
    Semillon, 74
    Trebbiano, 42

Hanford, California, 86
Harvest Festival, Livermore, 80
Hoff, Tracey, 79
Hogan, Francis 164
Home Ranch vineyard, 45, 69–76, 108
Heubner, Max, 103–104

Imperial Dynasty restaurant, 86–87
Ivan Támas Winery, 41–43, 79

Judgment of Paris tasting (1976), 19, 34, 61

"Klondike Kate" (Kate Bankowski), 69

L'Autre Côte Wines, 17, 62, 66
La Rochelle Pinot Noir, 78, 101, 106
Las Positas Vineyards, 17
Lawrence Livermore National Lab, 15
Leisure Street Winery, 17
Lineage Wine Company, 16, 17, 76, 130–131, 135–138
    pricing, 43–44, 172–173
    wine scores, 16
Livermore Valley, 5–6
    appellation, 14
    climate of, 62, 65–66
    history of winemaking in, 14–15, 74

similarity to Sauterne region of Bordeaux, 74
terroir, 6, 66
winemaking in, 15–20, 42–43, 62–63
Livermore, Robert, 14
Loma Prieta earthquake, 156
Longevity Wines, 17
Luce block, 152, 157, 158–159
Lynch, Kermit, 59
Lur-Saluces, 74

McGrail Vineyards & Winery, 17
Mel, Louis, 74
Mendenhall, William, 15
méthode champenoise, 73, 98–99
méthode ancestrale, 99
Mia Nipote Wines, 17
Mirassou Vineyards
 development of, 36–42
 sale to Gallo, 33, 106
 tasting room, 104–105
Mirassou, Aidan, 33, 34–36, 78, 117
Mirassou, Edmund Aloysius, 31, 98
Mirassou, June, 23–27, 78, 145–149
Mirassou, Justine Schreiber, 30
Mirassou, Katherine, 25
Mirassou, Norbert Charles, 31, 32, 97
Mirassou, Peter, 30
Mirassou, Henriette Pellier, 28, 29
Mirassou, Pierre, 29
Mirassou, Sara, 25, 44
Mirassou, Steve (father), 32, 37–39, 86–91, 128
Mirassou, Steven Kent
 children, 25
 education, 24
 developing vintages, 91–94
 family history, 28–36
 marriage, 25
 move to Livermore, 25
 Thanksgiving, 107–119
 training, 97–106

wine palate, 130–131, 135–136
wine-making philosophy, 37, 176–177
Mission San Jose, 14
Mitchell Katz Winery, 17
Mondavi, Robert, 38, 40
Monterey County, Mirassou in, 40–41, 106
Mount Diablo Highlands Wine Quality Alliance, 17
Mount Madonna, 152
Mourvèdre, 34
Murrieta's Well, 17

Napa Valley, climate, 73
Napa Valley, vs. Livermore Valley, 18–19
Nella Terra Cellars, 17
Ness, Laura, 154
Ninja Angel Cabernet and Merlot blend, 149
Nottingham Cellars, 17

Ohlone people, 14
Omega Road Winery, 17
Out from Land, 145–149, 165–167

Page Mill Winery, 17
Paris tasting, 1976, 19, 34
Parker, Robert, 19
Patterson, Jeffrey, 164
Pellier, Henriette, 28
Pellier, Pierre Renaud, 28, 106
Pétillant Naturel, 99
Pepper Tree Horse Farm, 75
Ploof, Craig, 77–79
Pinot Grigio, 42
Pinot Noir, 34, 40, 65, 78, 99, 101, 104, 106, 123, 154, 176
Pol Roger Brut Chardonnay, 116
Positive/Negative Cabernet Sauvignon, 45
Premier, The, Cabernet Sauvignon, 76
pyrazines, 63, 64

Rancho Philo sherry

(Cucamonga), 136
Ramirez, Tracey, 70
Refsnider, Beth, 35, 171
Refsnider, Patrick, 171
Retzlaff Vineyards & Estate Winery, 17
restaurants, 86–90
Ridge Trail block, 152, 157
Rios-Lovell Winery, 17
Rodrigue-Molyneaux Winery, 18
Rosa Fierro Cellars, 18
Ruby Hill Winery, 18

Sachau Vineyard, 75, 124, 171–172
San Jose, 24, 97
Sandia Labs, 15
Sangiovese, 42, 50, 71, 113, 171
Santa Cruz Mountain, 153
Santa Cruz Mountain Vineyards, 154–156
Sauvignon Blanc, 6, 41, 61, 63, 74, 75, 171
Schoendienst, Kathy, 79
Semillon, 74
Semillon-Sauvignon Blanc, 15
sherry, Rancho Philo, 136
Sideways (movie), 106
Singing Winemaker, 18
sparkling wines, 73, 98–101, 113
Spurrier, Stephen, 61
Stagecoach block, 152, 157
Steven Kent Winery, 5, 16, 17, 39, 42, 79, 167–168
 founding of, 41
 pricing of wines, 43–44, 172–173
 tasting room, 70–71
 wine scores, 16
Stutz, Tom, 106

terroir
 and 2017 vintage, 135
 Bates Ranch, 152–153, 156
 Home Ranch, 71, 75
 Livermore Valley, 6, 66
3 Steves Winery, 17
Toomey, Matt, 171

Toomey, Rebecca, 171
Trebbiano, 42
Tri-Valley Conservancy, 16
Troupe-Masi, Jeremy, 164

University of California, Davis, 61

Vincere super Tuscan, 72

Wente Vineyards, 16, 18, 42, 79, 139
Wente, Phil, 42
Wetmore, Charles, 15, 74
wine clubs, 45, 114
Wine Enthusiast, 128
Wine Spectator, 19, 128
wine
   actualizing true character, 123
   age of, 127–130
   emotional reactions to, 135–138
   food pairings, 85–90, 171–174
   flavor notes, 53–55, 74, 89, 100, 109, 122, 125, 137
   mystical properties of, 12
   pricing, 43–44, 172–173
   tasting, 13, 154–156
   vintages, 127, 135–138

wine-making
   barrel-aging, 110–111
   biodynamic, 125, 132–133
   climate change and, 128
   controlling brix, 51, 64
   craft of, 176–177
   data-driven, 141
   development of vintages, 8, 91–94
   process, 47–57, 123
   fermentation, 47, 48–56
   grape crushing, 56–57
   grape growing, 4–5, 7, 172, 175–176
   grape harvest, 142
   maceration, 53
   punching down, 53

winemakers, 139–142

Claude Bobba, 139–142
Max Heubner, 103–104
Aidan Mirassou (assistant), 33, 47, 91
Craig Ploof (assistant), 77–79
Beth Refsnider (assistant), 35, 47, 91
Tom Stutz, 106
winery workers, 4, 101
winese, 103
Wing, Richard, 86–88
Wood Family Vineyards, 18

yeast, 48–50, 100
yeast, *Brettanomyces*, 124–125